BRITANNICA

Mathematics in Context

Statistics and the Environment

Britannica

ENCYCLOPÆDIA BRITANNICA EDUCATIONAL CORPORATION

Mathematics in Context is a comprehensive curriculum for the middle grades. It was developed in collaboration with the Wisconsin Center for Education Research, School of Education, University of Wisconsin–Madison and the Freudenthal Institute at the University of Utrecht, The Netherlands, with the support of National Science Foundation Grant No. 9054928.

National Science Foundation
Opinions expressed are those of the authors and not necessarily those of the Foundation

ISBN 0-7826-1537-6
1 2 3 4 5 WK 02 01 00 99 98

The *Mathematics in Context* Development Team

Mathematics in Context is a comprehensive curriculum for the middle grades. The National Science Foundation funded the National Center for Research in Mathematical Sciences Education at the University of Wisconsin–Madison to develop and field-test the materials from 1991 through 1996. The Freudenthal Institute at the University of Utrecht in The Netherlands, as a subcontractor, collaborated with the University of Wisconsin–Madison on the development of the curriculum.

The initial version of *Statistics and the Environment* was developed by Vincent Jonker, Nanda Querelle, Monica Wijers, and Astrid de Wild. It was adapted for use in American schools by Mary S. Spence, Mary Ann Fix, Mary C. Shafer, and James Browne.

National Center for Research in Mathematical Sciences Education Staff

Thomas A. Romberg
Director

Joan Daniels Pedro
Assistant to the Director

Gail Burrill
Coordinator
Field Test Materials

Margaret R. Meyer
Coordinator
Pilot Test Materials

Mary Ann Fix
Editorial Coordinator

Sherian Foster
Editorial Coordinator

James A. Middleton
Pilot Test Coordinator

Margaret A. Pligge
First Edition Coordinator

Project Staff

Jonathan Brendefur
Laura J. Brinker
James Browne
Jack Burrill
Rose Byrd
Peter Christiansen
Barbara Clarke
Doug Clarke
Beth R. Cole
Fae Dremock
Jasmina Milinkovic
Kay Schultz
Mary C. Shafer
Julia A. Shew
Aaron N. Simon
Marvin Smith
Stephanie Z. Smith
Mary S. Spence
Kathleen A. Steele

Freudenthal Institute Staff

Jan de Lange
Director

Els Feijs
Coordinator

Martin van Reeuwijk
Coordinator

Project Staff

Mieke Abels
Nina Boswinkel
Frans van Galen
Koeno Gravemeijer
Marja van den Heuvel-Panhuizen
Jan Auke de Jong
Vincent Jonker
Ronald Keijzer
Martin Kindt
Jansie Niehaus
Nanda Querelle
Anton Roodhardt
Leen Streefland
Adri Treffers
Monica Wijers
Astrid de Wild

Acknowledgments

Several school districts used and evaluated one or more versions of the materials: Ames Community School District, Ames, Iowa; Parkway School District, Chesterfield, Missouri; Stoughton Area School District, Stoughton, Wisconsin; Madison Metropolitan School District, Madison, Wisconsin; Milwaukee Public Schools, Milwaukee, Wisconsin; and Dodgeville School District, Dodgeville, Wisconsin. Two sites were involved in staff development as well as formative evaluation of materials: Culver City, California, and Memphis, Tennessee. Two sites were developed through partnership with Encyclopædia Britannica Educational Corporation: Miami, Florida, and Puerto Rico. University Partnerships were developed with mathematics educators who worked with preservice teachers to familiarize them with the curriculum and to obtain their advice on the curriculum materials. The materials were also used at several other schools throughout the United States.

We at Encyclopædia Britannica Educational Corporation extend our thanks to all who had a part in making this program a success. Some of the participants instrumental in the program's development are as follows:

Allapattah Middle School
Miami, Florida
Nemtalla (Nikolai) Barakat

Ames Middle School
Ames, Iowa
Kathleen Coe
Judd Freeman
Gary W. Schnieder
Ronald H. Stromen
Lyn Terrill

Bellerive Elementary
Creve Coeur, Missouri
Judy Hetterscheidt
Donna Lohman
Gary Alan Nunn
Jakke Tchang

Brookline Public Schools
Brookline, Massachusetts
Rhonda K. Weinstein
Deborah Winkler

Cass Middle School
Milwaukee, Wisconsin
Tami Molenda
Kyle F. Witty

Central Middle School
Waukesha, Wisconsin
Nancy Reese

Craigmont Middle School
Memphis, Tennessee
Sharon G. Ritz
Mardest K. VanHooks

Crestwood Elementary
Madison, Wisconsin
Diane Hein
John Kalson

Culver City Middle School
Culver City, California
Marilyn Culbertson
Joel Evans
Joy Ellen Kitzmiller
Patricia R. O'Connor
Myrna Ann Perks, Ph.D.
David H. Sanchez
John Tobias
Kelley Wilcox

Cutler Ridge Middle School
Miami, Florida
Lorraine A. Valladares

Dodgeville Middle School
Dodgeville, Wisconsin
Jacqueline A. Kamps
Carol Wolf

Edwards Elementary
Ames, Iowa
Diana Schmidt

Fox Prairie Elementary
Stoughton, Wisconsin
Tony Hjelle

Grahamwood Elementary
Memphis, Tennessee
M. Lynn McGoff
Alberta Sullivan

Henry M. Flagler Elementary
Miami, Florida
Frances R. Harmon

Horning Middle School
Waukesha, Wisconsin
Connie J. Marose
Thomas F. Clark

Huegel Elementary
Madison, Wisconsin
Nancy Brill
Teri Hedges
Carol Murphy

Hutchison Middle School
Memphis, Tennessee
Maria M. Burke
Vicki Fisher
Nancy D. Robinson

Idlewild Elementary
Memphis, Tennessee
Linda Eller

Jefferson Elementary
Santa Ana, California
Lydia Romero-Cruz

Jefferson Middle School
Madison, Wisconsin
Jane A. Beebe
Catherine Buege
Linda Grimmer
John Grueneberg
Nancy Howard
Annette Porter
Stephen H. Sprague
Dan Takkunen
Michael J. Vena

Jesus Sanabria Cruz School
Yabucoa, Puerto Rico
Andreíta Santiago Serrano

John Muir Elementary School
Madison, Wisconsin
Julie D'Onofrio
Jane M. Allen-Jauch
Kent Wells

Kegonsa Elementary
Stoughton, Wisconsin
Mary Buchholz
Louisa Havlik
Joan Olsen
Dominic Weisse

Linwood Howe Elementary
Culver City, California
Sandra Checel
Ellen Thireos

Mitchell Elementary
Ames, Iowa
Henry Gray
Matt Ludwig

New School of Northern Virginia
Fairfax, Virginia
Denise Jones

Northwood Elementary
Ames, Iowa
Eleanor M. Thomas

Orchard Ridge Elementary
Madison, Wisconsin
Mary Paquette
Carrie Valentine

Parkway West Middle School
Chesterfield, Missouri
Elissa Aiken
Ann Brenner
Gail R. Smith

Ridgeway Elementary
Ridgeway, Wisconsin
Lois Powell
Florence M. Wasley

Roosevelt Elementary
Ames, Iowa
Linda A. Carver

Roosevelt Middle
Milwaukee, Wisconsin
Sandra Simmons

Ross Elementary
Creve Coeur, Missouri
Annette Isselhard
Sheldon B. Korklan
Victoria Linn
Kathy Stamer

St. Joseph's School
Dodgeville, Wisconsin
Rita Van Dyck
Sharon Wimer

St. Maarten Academy
St. Peters, St. Maarten, NA
Shareed Hussain

Sarah Scott Middle School
Milwaukee, Wisconsin
Kevin Haddon

Sawyer Elementary
Ames, Iowa
Karen Bush Hoiberg

Sennett Middle School
Madison, Wisconsin
Brenda Abitz
Lois Bell
Shawn M. Jacobs

Sholes Middle School
Milwaukee, Wisconsin
Chris Gardner
Ken Haddon

Stephens Elementary
Madison, Wisconsin
Katherine Hogan
Shirley M. Steinbach
Kathleen H. Vegter

Stoughton Middle School
Stoughton, Wisconsin
Sally Bertelson
Polly Goepfert
Jacqueline M. Harris
Penny Vodak

Toki Middle School
Madison, Wisconsin
Gail J. Anderson
Vicky Grice
Mary M. Ihlenfeldt
Steve Jernegan
Jim Leidel
Theresa Loehr
Maryann Stephenson
Barbara Takkunen
Carol Welsch

Trowbridge Elementary
Milwaukee, Wisconsin
Jacqueline A. Nowak

W. R. Thomas Middle School
Miami, Florida
Michael Paloger

Wooddale Elementary Middle School
Memphis, Tennessee
Velma Quinn Hodges
Jacqueline Marie Hunt

Yahara Elementary
Stoughton, Wisconsin
Mary Bennett
Kevin Wright

Site Coordinators

Mary L. Delagardelle—Ames Community Schools, Ames, Iowa

Dr. Hector Hirigoyen—Miami, Florida

Audrey Jackson—Parkway School District, Chesterfield, Missouri

Jorge M. López—Puerto Rico

Susan Militello—Memphis, Tennessee

Carol Pudlin—Culver City, California

Reviewers and Consultants

Michael N. Bleicher
Professor of Mathematics
University of Wisconsin–Madison
Madison, WI

Diane J. Briars
Mathematics Specialist
Pittsburgh Public Schools
Pittsburgh, PA

Donald Chambers
Director of Dissemination
University of Wisconsin–Madison
Madison, WI

Don W. Collins
Assistant Professor of Mathematics Education
Western Kentucky University
Bowling Green, KY

Joan Elder
Mathematics Consultant
Los Angeles Unified School District
Los Angeles, CA

Elizabeth Fennema
Professor of Curriculum and Instruction
University of Wisconsin-Madison
Madison, WI

Nancy N. Gates
University of Memphis
Memphis, TN

Jane Donnelly Gawronski
Superintendent
Escondido Union High School
Escondido, CA

M. Elizabeth Graue
Assistant Professor of Curriculum and Instruction
University of Wisconsin–Madison
Madison, WI

Jodean E. Grunow
Consultant
Wisconsin Department of Public Instruction
Madison, WI

John G. Harvey
Professor of Mathematics and Curriculum & Instruction
University of Wisconsin–Madison
Madison, WI

Simon Hellerstein
Professor of Mathematics
University of Wisconsin–Madison
Madison, WI

Elaine J. Hutchinson
Senior Lecturer
University of Wisconsin–Stevens Point
Stevens Point, WI

Richard A. Johnson
Professor of Statistics
University of Wisconsin–Madison
Madison, WI

James J. Kaput
Professor of Mathematics
University of Massachusetts–Dartmouth
Dartmouth, MA

Richard Lehrer
Professor of Educational Psychology
University of Wisconsin–Madison
Madison, WI

Richard Lesh
Professor of Mathematics
University of Massachusetts–Dartmouth
Dartmouth, MA

Mary M. Lindquist
Callaway Professor of Mathematics Education
Columbus College
Columbus, GA

Baudilio (Bob) Mora
Coordinator of Mathematics & Instructional Technology
Carrollton-Farmers Branch Independent School District
Carrollton, TX

Paul Trafton
Professor of Mathematics
University of Northern Iowa
Cedar Falls, IA

Norman L. Webb
Research Scientist
University of Wisconsin–Madison
Madison, WI

Paul H. Williams
Professor of Plant Pathology
University of Wisconsin–Madison
Madison, WI

Linda Dager Wilson
Assistant Professor
University of Delaware
Newark, DE

Robert L. Wilson
Professor of Mathematics
University of Wisconsin–Madison
Madison, WI

TABLE OF CONTENTS

Letter to the Teacher . . . viii

Overview

How to Use This Book . . . x
Unit Focus . . . xii
Planning and Preparation . . . xiii
Assessment . . . xv
Connections across the *Mathematics in Context* Curriculum . . . xix

Student Materials and Teaching Notes

Student Book Table of Contents . . . 2
Letter to the Student . . . 3

Section A Preparing to Use Statistics

Section Overview . . . 4
Murre Island . . . 6
Weather . . . 10
Murre Island Bats . . . 14
Topography . . . 16
Describing Murre Island . . . 24
Collecting Data . . . 28
Summary . . . 32

Section B Data about Endangered Forests

Section Overview . . . 34
Murre Island Forests . . . 36
National Forests . . . 42
Summary . . . 46

Section C Interpreting Data

Section Overview . . . 48
Statistical Summaries . . . 50
Vacation Activities . . . 54
Summary . . . 60

Section D Describing Data

Section Overview . . . 62
Environmental Problems . . . 64
Water . . . 66
Energy . . . 72
Being Economical with Electricity . . . 78
Energy Use on Murre Island . . . 78
Waste Disposal . . . 80
Waste Disposal on Murre Island . . . 90
Summary . . . 90

Section E Land-Use Plans for Murre Island

Section Overview . . . 92
Campgrounds . . . 100
Hiking Trails . . . 102
Lodge . . . 104
Recreation . . . 106

Assessment Activities and Solutions

Assessment Overview . . . 108
Garbage . . . 110
Lake Superior Island . . . 114
Hiking on an Island . . . 116

Glossary

Statistics and the Environment Glossary . . . 118

Blackline Masters

Letter to the Family . . . 120
Student Activity Sheets . . . 121
Assessment Activities . . . 130

Credits . . . 134

Dear Teacher,

Welcome! *Mathematics in Context* is designed to reflect the National Council of Teachers of Mathematics Standards for School Mathematics and to ground mathematical content in a variety of real-world contexts. Rather than relying on you to explain and demonstrate generalized definitions, rules, or algorithms, students investigate questions directly related to a particular context and construct mathematical understanding and meaning from that context.

The curriculum encompasses 10 units per grade level. *Statistics and the Environment* is designed to be the fourth unit in the statistics strand, after the grade 5/6 unit *Picturing Numbers* and the grade 6/7 unit *Dealing with Data.* The unit also lends itself to independent use—to enrich students' understanding of gathering, interpreting, and presenting data.

In addition to the Teacher Guide and Student Books, *Mathematics in Context* offers the following components that will inform and support your teaching:

- *Teacher Resource and Implementation Guide,* which provides an overview of the complete system, including program implementation, philosophy, and rationale
- *Number Tools,* Volumes 1 and 2, which are a series of blackline masters that serve as review sheets or practice pages involving number issues and basic skills
- *News in Numbers,* which is a set of additional activities that can be inserted between or within other units; it includes a number of measurement problems that require estimation.

Thank you for choosing *Mathematics in Context.* We wish you success and inspiration!

Sincerely,

The Mathematics in Context Development Team

Number Strand | Algebra Strand | Geometry Strand | Statistics Strand

Overview

24 cm
?
BRITANNICA
Mathematics in Context
Year
World Population in Billions
1630
0.5
1820
1.0
1890
1.5
1930
2.0
1950
2.5
1960
3.0
1968
3.5
1975
4.0
1981
4.5
1987
5.0
FROG NEWTON
Portland
Eugene
Bend
Grants Pass
Klamath Falls
FAVORITE SPORTS

How to Use This Book

This unit is one of 40 for the middle grades. Each unit can be used independently; however, the 40 units are designed to make up a complete, connected curriculum (10 units per grade level). There is a Student Book and a Teacher Guide for each unit.

Each Teacher Guide comprises elements that assist the teacher in the presentation of concepts and in understanding the general direction of the unit and the program as a whole. Becoming familiar with this structure will make using the units easier.

Each Teacher Guide consists of six basic parts:

- Overview
- Student Materials and Teaching Notes
- Assessment Activities and Solutions
- Glossary
- Blackline Masters
- Try This! Solutions

Overview

Before beginning this unit, read the Overview in order to understand the purpose of the unit and to develop strategies for facilitating instruction. The Overview provides helpful information about the unit's focus, pacing, goals, and assessment, as well as explanations about how the unit fits with the rest of the *Mathematics in Context* curriculum.

Student Material and Teaching Notes

This Teacher Guide contains all of the student pages, (except the Try This! activities), each of which faces a page of solutions, samples of students' work, and hints and comments about how to facilitate instruction. Note: Solutions for the Try This! activities can be found at the back of the Teacher Guide.

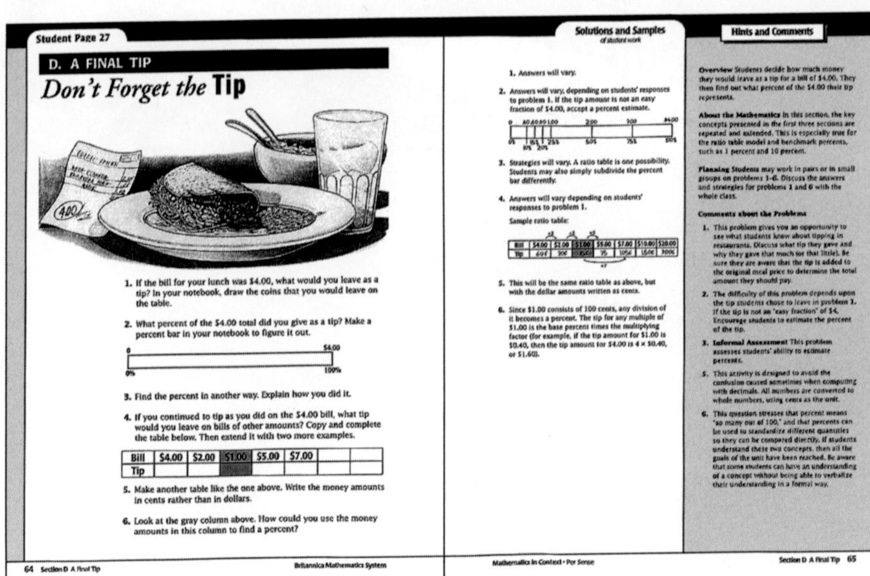

Student Page 27

D. A FINAL TIP

Don't Forget the **Tip**

1. If the bill for your lunch was $4.00, what would you leave as a tip? In your notebook, draw the coins that you would leave on the table.
2. What percent of the $4.00 total did you give as a tip? Make a percent bar in your notebook to figure it out.
3. Find the percent in another way. Explain how you did it.
4. If you continued to tip as you did on the $4.00 bill, what tip would you leave on bills of other amounts? Copy and complete the table below. Then extend it with two more examples.

Bill	$4.00	$2.00	$1.00	$5.00	$7.00		
Tip							

5. Make another table like the one above. Write the money amounts in cents rather than in dollars.
6. Look at the gray column above. How could you use the money amounts in this column to find a percent?

64 Section D A Final Tip — Britannica Mathematics System

Solutions and Samples *of student work*

1. Answers will vary.
2. Answers will vary, depending on students' responses to problem 1. If the tip amount is not an easy fraction of $4.00, accept a percent estimate.
3. Strategies will vary. A ratio table is one possibility. Students may also simply subdivide the percent bar differently.
4. Answers will vary depending on students' responses to problem 1.

Sample ratio table:

Bill	$4.00	$2.00	$1.00	$5.00	$7.00	$10.00	$20.00
Tip	60¢	30¢	15¢	75	105¢	150¢	300¢

5. This will be the same ratio table as above, but with the dollar amounts written as cents.
6. Since $1.00 consists of 100 cents, any division of it becomes a percent. The tip for any multiple of $1.00 is the base percent times the multiplying factor (for example, if the tip amount for $1.00 is $0.40, then the tip amount for $4.00 is 4 × $0.40, or $1.60).

Hints and Comments

Overview Students decide how much money they would leave as a tip for a bill of $4.00. They then find out what percent of the $4.00 their tip represents.

About the Mathematics In this section, the key concepts presented in the first three sections are repeated and extended. This is especially true for the ratio table model and benchmark percents, such as 1 percent and 10 percent.

Planning Students may work in pairs or in small groups on problems 1–6. Discuss the answers and strategies for problems 1 and 6 with the whole class.

Comments about the Problems

1. This problem gives you an opportunity to see what students know about tipping in restaurants. Discuss what tip they gave and why they gave that much (or that little). Be sure they are aware that the tip is added to the original meal price to determine the total amount they should pay.
2. The difficulty of this problem depends upon the tip students chose to leave in problem 1. If the tip is not an "easy fraction" of $4, Encourage students to estimate the percent of the tip.
3. **Informal Assessment** This problem assesses students' ability to estimate percents.
5. This activity is designed to avoid the confusion caused sometimes when computing with decimals. All numbers are converted to whole numbers, using cents as the unit.
6. This question stresses that percent means "so many out of 100," and that percents can be used to standardize different quantities so they can be compared directly. If students understand these two concepts, then all the goals of the unit have been reached. Be aware that some students can have an understanding of a concept without being able to verbalize their understanding in a formal way.

Mathematics in Context • Per Sense — Section D A Final Tip 65

Each section within the unit begins with a two-page spread that describes the work students do, the goals of the section, new vocabulary, and materials needed, as well as providing information about the mathematics in the section and ideas for pacing, planning instruction, homework, and assessment.

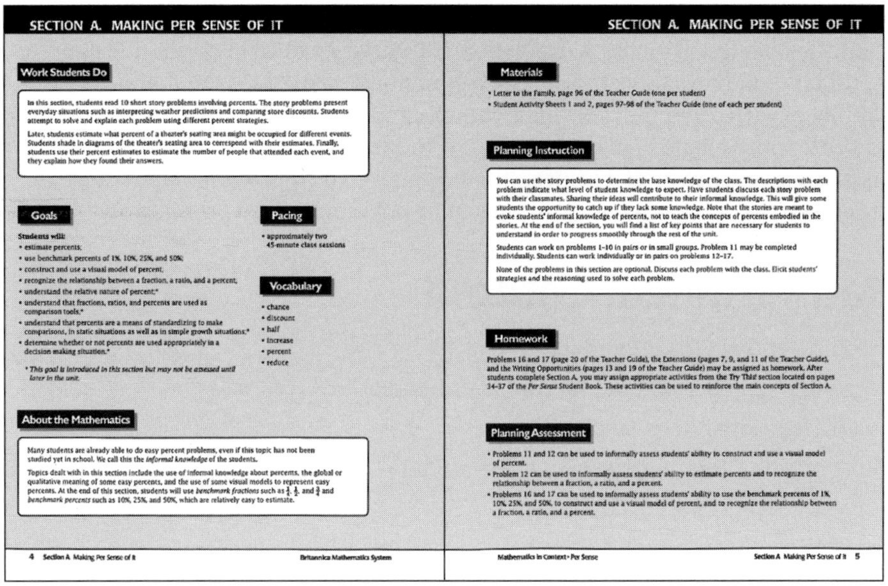

SECTION A. MAKING PER SENSE OF IT

Work Students Do

In this section, students read 10 short story problems involving percents. The story problems present everyday situations such as interpreting weather predictions and comparing store discounts. Students attempt to solve and explain each problem using different percent strategies.

Later, students estimate what percent of a theater's seating area might be occupied for different events. Students shade in diagrams of the theater's seating area to correspond with their estimates. Finally, students use their percent estimates to estimate the number of people that attended each event, and they explain how they found their answers.

Goals

Students will:
- estimate percents;
- use benchmark percents of 1%, 10%, 25%, and 50%;
- construct and use a visual model of percent;
- recognize the relationship between a fraction, a ratio, and a percent;
- understand the relative nature of percent;*
- understand that fractions, ratios, and percents are used as comparison tools;*
- understand that percents are a means of standardizing to make comparisons, in static situations as well as in simple growth situations;*
- determine whether or not percents are used appropriately in a decision making situation.*

** This goal is introduced in this section but may not be assessed until later in the unit.*

Pacing

- approximately two 45-minute class sessions

Vocabulary

- chance
- discount
- half
- increase
- percent
- reduce

About the Mathematics

Many students are already able to do easy percent problems, even if this topic has not been studied yet in school. We call this the *informal knowledge* of the students.

Topics dealt with in this section include the use of informal knowledge about percents, the global or qualitative meaning of some easy percents, and the use of some visual models to represent easy percents. At the end of this section, students will use *benchmark fractions* such as $\frac{1}{4}$, $\frac{1}{2}$, and $\frac{3}{4}$ and *benchmark percents* such as 10%, 25%, and 50%, which are relatively easy to estimate.

4 Section A Making Per Sense of It Britannica Mathematics System

SECTION A. MAKING PER SENSE OF IT

Materials

- Letter to the Family, page 96 of the Teacher Guide (one per student)
- Student Activity Sheets 1 and 2, pages 97–98 of the Teacher Guide (one of each per student)

Planning Instruction

You can use the story problems to determine the base knowledge of the class. The descriptions with each problem indicate what level of student knowledge to expect. Have students discuss each story problem with their classmates. Sharing their ideas will contribute to their informal knowledge. This will give some students the opportunity to catch up if they lack some knowledge. Note that the stories are meant to evoke students' informal knowledge of percents, not to teach the concepts of percents embodied in the stories. At the end of the section, you will find a list of key points that are necessary for students to understand in order to progress smoothly through the rest of the unit.

Students can work on problems 1–10 in pairs or in small groups. Problem 11 may be completed individually. Students can work individually or in pairs on problems 12–17.

None of the problems in this section are optional. Discuss each problem with the class. Elicit students' strategies and the reasoning used to solve each problem.

Homework

Problems 16 and 17 (page 20 of the Teacher Guide), the Extensions (pages 7, 9, and 11 of the Teacher Guide), and the Writing Opportunities (pages 13 and 19 of the Teacher Guide) may be assigned as homework. After students complete Section A, you may assign appropriate activities from the Try This! section located on pages 34–37 of the *Per Sense* Student Book. These activities can be used to reinforce the main concepts of Section A.

Planning Assessment

- Problems 11 and 12 can be used to informally assess students' ability to construct and use a visual model of percent.
- Problem 12 can be used to informally assess students' ability to estimate percents and to recognize the relationship between a fraction, a ratio, and a percent.
- Problems 16 and 17 can be used to informally assess students' ability to use the benchmark percents of 1%, 10%, 25%, and 50%, to construct and use a visual model of percent, and to recognize the relationship between a fraction, a ratio, and a percent.

Mathematics in Context • Per Sense Section A Making Per Sense of It 5

Assessment Activities and Solutions

Information about assessment can be found in several places in this Teacher Guide. General information about assessment is given in the Overview; informal assessment opportunities are identified on the teacher pages that face each student page; and the Assessment Activities section of this guide provides formal assessment opportunities.

Glossary

The Glossary defines all vocabulary words listed on the Section Opener pages. It includes mathematical terms that may be new to students, as well as words associated with the contexts introduced in the unit. (Note: The Student Book does not have a glossary. This allows students to construct their own definitions, based on their personal experiences with the unit activities.)

Blackline Masters

At the back of this Teacher Guide are blackline masters for photocopying. The blackline masters include a letter to families (to be sent home with students before beginning the unit), several student activity sheets, and assessment masters.

Try This! Solutions

Also included in the back of this Teacher Guide are the solutions to several Try This! activities—one related to each section of the unit—that can be used to reinforce the unit's main concepts. The Try This! activities are located in the back of the Student Book.

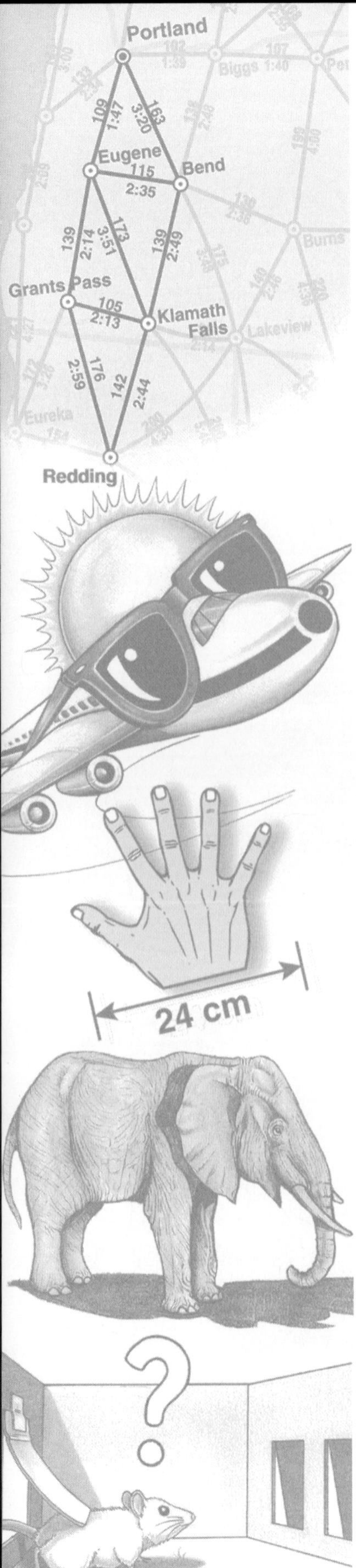

Unit Focus

Statistics and the Environment is an interdisciplinary unit that integrates science, language arts, social studies, and mathematics. This unit emphasizes statistical concepts, but mathematical concepts such as area, ratio, and proportion are also important. The emphasis is on statistical reasoning, on gaining knowledge from data. Students use statistics to help them make and defend land-use decisions for an environmentally sensitive island. The unit culminates with students working in small groups to design plans to preserve and protect the island. This unit is meant to provoke students to think and reason about controversial issues. It is not intended to promote or advance a particular viewpoint or support a particular position.

Mathematical Content

- reading and interpreting data
- organizing, describing, and representing data
- using fractions, decimals, and percents
- investigating the representativeness of data
- investigating random sampling
- designing and conducting a survey
- analyzing data and making inferences based on data
- using statistics to make decisions
- using and making area maps, contour maps, and cross sections (profiles)

Prior Knowledge

This unit assumes that students can do the following with understanding:

- add, subtract, multiply, and divide rational numbers in all representations (decimals, percentages, and fractions);
- read, interpret, make, and use graphical representations of data (scatter plots, box plots, number line plots, stem-and-leaf plots, and histograms) as introduced and used in the units *Picturing Numbers* and *Dealing with Data;*
- understand and use statistical measures (mean, median, and mode) as introduced and used in the units *Picturing Numbers* and *Dealing with Data;*
- create random samples as introduced in *Dealing with Data;*
- use metric measurements;
- use and make maps as introduced in the units *Grasping Sizes* and *Reallotment.*

This unit should be taught after the grade 5/6 unit *Picturing Numbers* and the grade 6/7 units *Dealing with Data* and *Reallotment.*

Planning and Preparation

Pacing: 17 days

Section	Work Students Do	Pacing*	Materials
A. Preparing to Use Statistics	■ read and interpret graphical and numerical data ■ design and conduct a survey ■ investigate the representativeness of data ■ make inferences based on data	4 days	■ Letter to the Family (one per student) ■ Student Activity Sheets 1–4 (one of each per student) ■ See page 5 of the Teacher Guide for a complete list of the materials and quantities needed.
B. Data about Endangered Forests	■ read and interpret graphical and numerical data ■ use fractions, decimals, and percents ■ organize and describe data ■ make inferences based on data	3 days	■ Student Activity Sheets 1, 6, and 9 (one of each per student) ■ Student Activity Sheets 7 and 8, optional (one of each per student) ■ See page 35 of the Teacher Guide for a complete list of the materials and quantities needed.
C. Interpreting Data	■ organize and describe data ■ use fractions, decimals, and percents ■ represent and analyze data ■ investigate the representativeness of data ■ investigate random sampling ■ make inferences based on data ■ use statistics to make decisions	4 days	■ See page 49 of the Teacher Guide for a complete list of the materials and quantities needed.
D. Describing Data	■ read and interpret graphs ■ use fractions, decimals, and percents ■ organize and describe data ■ represent data ■ make inferences based on data ■ use statistics to make decisions	3 days	■ graph paper (seven sheets per student) ■ calculators (one per student)
E. Land-Use Plans for Murre Island	■ assess the use of statistical summaries and graphs; read, interpret, and use information from maps or graphical displays of data; and make inferences and decisions based on data and statistical information and justify these decisions.	2 days	■ Student Activity Sheets 1 and 2, optional (one of each per student) ■ See page 93 of the Teacher Guide for a complete list of the materials and quantities needed.

* One day is approximately equivalent to one 45-minute class session.

Preparation

In the *Teacher Resource and Implementation Guide* is an extensive description of the philosophy underlying both the content and the pedagogy of the *Mathematics in Context* curriculum. Suggestions for preparation are also given in the Hints and Comments columns of this Teacher Guide. You may want to consider the following:

- Work through the unit before teaching it. If possible, take on the role of the student and discuss your strategies with other teachers.
- Use the overhead projector for student demonstrations, particularly with overhead transparencies of the student activity sheets and any manipulatives used in the unit.
- Invite students to use drawings and examples to illustrate and clarify their answers.
- Allow students to work at different levels of sophistication. Some students may need concrete materials, while others can work at a more abstract level.
- Provide opportunities and support for students to share their strategies, which often differ. This allows students to take part in class discussions and introduces them to alternative ways to think about the mathematics in the unit.
- In some cases, it may be necessary to read the problems to students or to pair students to facilitate their understanding of the printed materials.
- A list of the materials needed for this unit is in the chart on page xiii.
- Try to follow the recommended pacing chart on page xiii. You can easily spend more time on this unit than the number of class periods indicated. Bear in mind, however, that many of the topics introduced in this unit will be revisited and covered more thoroughly in other *Mathematics in Context* units.

Resources

For Teachers

Books and Magazines

- *Mathematics Assessment: Myths, Models, Good Questions, and Practical Suggestions,* edited by Jean Kerr Stenmark (Reston, Virginia: The National Council of Teachers of Mathematics, Inc., 1991)
- *Birds of North America: An Audubon Society Beginner's Guide,* by George S. Fichter (New York: Random House, 1982)
- *Batman: Exploring the World of Bats,* by Laurence Pringle (New York: Scribner, 1991)

For Students

Videos for Students

Mathsphere Videos

- *Finding the Way*
- *What's the Connection*
- *An Area of Concern*
- *A Need for Data*
- *Fields of Plenty*
- *An Exhausting Quest*
- *A Value for Cleanliness*

(available from Encyclopædia Britannica)

Assessment

Planning Assessment

In keeping with the NCTM Assessment Standards, valid assessment should be based on evidence drawn from several sources. (See the full discussion of assessment philosophies in the *Teacher Resource and Implementation Guide.*) An assessment plan for this unit may draw from the following sources:

- Observations—look, listen, and record observable behavior.
- Interactive Responses—in a teacher-facilitated situation, note how students respond, clarify, revise, and extend their thinking.
- Products—look for the quality of thought evident in student projects, test answers, worksheet solutions, or writings.

These categories are not meant to be mutually exclusive. In fact, observation is a key part of assessing interactive responses and also key to understanding the end results of projects and writings.

Ongoing Assessment Opportunities

- **Problems within Sections**
  To evaluate ongoing progress, *Mathematics in Context* identifies informal assessment opportunities and the goals that these particular problems assess throughout the Teacher Guide. There are also indications as to what you might expect from your students.

  The land-use development projects in Section E can be used as a culminating project assessment at the end of the unit. See the Section E overview on page 93 of the Teacher Guide for additional information.

- **Section Summary Questions**
  The summary questions at the end of each section are vehicles for informal assessment (see Teacher Guide pages 32, 46, 60, and 90).

End-of-Unit Assessment Opportunities

In the back of this Teacher Guide, there are three assessment activities that, when combined, require one to two 45-minute class sessions. For a more detailed description of these assessment activities, see the Assessment Overview (Teacher Guide pages 108 and 109).

You may also wish to design your own culminating project or let students create one that will tell you what they consider is important in the unit. For more assessment ideas, refer to the chart on pages xvi and xvii.

Goals and Assessment

In the *Mathematics in Context* curriculum, unit goals, categorized according to cognitive procedures, relate to the strand goals and the NCTM Curriculum and Evaluation Standards. Additional information about these goals is found in the *Teacher Resource and Implementation Guide.* The *Mathematics in Context* curriculum is designed to help students develop their abilities so that they can perform with understanding in each of the categories listed below. It is important to note that the attainment of goals in one category is not a prerequisite to attaining those in another category. In fact, students should progress simultaneously toward several goals in different categories.

	Goal	Ongoing Assessment Opportunities	End-of-Unit Assessment Opportunities
Conceptual and Procedural Knowledge	**1.** organize and describe data	**Section C** p. 50, #1 p. 54, #11 **Section E** pp. 100–106, all four activities	Garbage, pp. 130–131 Lake Superior Island, p. 132
	2. use statistical summaries and graphs	**Section C** p. 50, #1 p. 56, #13 **Section D** p. 68, #6 p. 72, #15 p. 74, #18 p. 86, #38 **Section E** pp. 100–106, all four activities	Garbage, pp. 130–131
	3. calculate with fractions, decimals, and percents	**Section B** p. 36, #2 **Section C** p. 54, #10 p. 60, #16, #17 **Section D** p. 74, #17 p. 84, #35–#37	Garbage, pp. 130–131

	Goal	Ongoing Assessment Opportunities	End-of-Unit Assessment Opportunities
Reasoning, Communicating, Thinking, and Making Connections	**4.** judge whether or not samples are representative and random	**Section C** p. 52, #4b, #5 p. 58, #14, #15	Lake Superior Island, p. 132
	5. read, interpret, and use information from maps or graphical displays of data	**Section A** p. 22, #18, #19 p. 24, #20 **Section B** p. 42, #11 **Section E** pp. 100–106, all four activities	Hiking on an Island, p. 133

	Goal	Ongoing Assessment Opportunities	End-of-Unit Assessment Opportunities
Modeling, Nonroutine Problem-Solving, Critically Analyzing, and Generalizing	**6.** choose and use appropriate techniques to analyze data	**Section C** p. 54, #11 **Section D** p. 68, #6 p. 74, #18 p. 86, #38	Garbage, pp. 130–131
	7. make inferences based on data	**Section C** p. 54, #11 p. 56, #13 **Section E** pp. 100–106, all four activities	Garbage, pp. 130–131 Lake Superior Island, p. 132
	8. make decisions based on statistical information and justify the decisions	**Section E** pp. 100–106, all four activities	Garbage, pp. 130–131

More about Assessment

Scoring and Analyzing Assessment Responses

Students may respond to assessment questions with various levels of mathematical sophistication and elaboration. Each student's response should be considered for the mathematics that it shows, and not judged on whether or not it includes an expected response. Responses to some of the assessment questions may be viewed as either correct or incorrect, but many answers will need flexible judgment by the teacher. Descriptive judgments related to specific goals and partial credit often provide more helpful feedback than percentage scores.

Openly communicate your expectations to all students, and report achievement and progress for each student relative to those expectations. When scoring students' responses try to think about how they are progressing toward the goals of the unit and the strand.

Student Portfolios

Generally, a portfolio is a collection of student-selected pieces that is representative of a student's work. A portfolio may include evaluative comments by you or by the student. See the *Teacher Resource and Implementation Guide* for more ideas on portfolio focus and use.

A comprehensive discussion about the contents, management, and evaluation of portfolios can be found in *Mathematics Assessment: Myths, Models, Good Questions, and Practical Suggestions*, pp. 35–48.

Student Self-Evaluation

Self-evaluation encourages students to reflect on their progress in learning mathematical concepts, their developing abilities to use mathematics, and their dispositions toward mathematics. The following examples illustrate ways to incorporate student self-evaluations as one component of your assessment plan.

- Ask students to comment, in writing, on each piece they have chosen for their portfolios and on the progress they see in the pieces overall.
- Give a writing assignment entitled "What I Know Now about [a math concept] and What I Think about It." This will give you information about each student's disposition toward mathematics as well as his or her knowledge.
- Interview individuals or small groups to elicit what they have learned, what they think is important, and why.

Suggestions for self-inventories can be found in *Mathematics Assessment: Myths, Models, Good Questions, and Practical Suggestions*, pp. 55–58.

Summary Discussion

Discuss specific lessons and activities in the unit—what the student learned from them and what the activities have in common. This can be done in whole-class discussions, small groups, or in personal interviews.

Connections across the *Mathematics in Context* Curriculum

Statistics and the Environment is the second unit in the statistics strand at the grade 7/8 level. The map below shows the complete *Mathematics in Context* curriculum for grade 7/8. It indicates where the unit fits in the statistics strand and where it fits in the overall picture.

A detailed description of the units, the strands, and the connections in the *Mathematics in Context* curriculum can be found in the *Teacher Resource and Implementation Guide.*

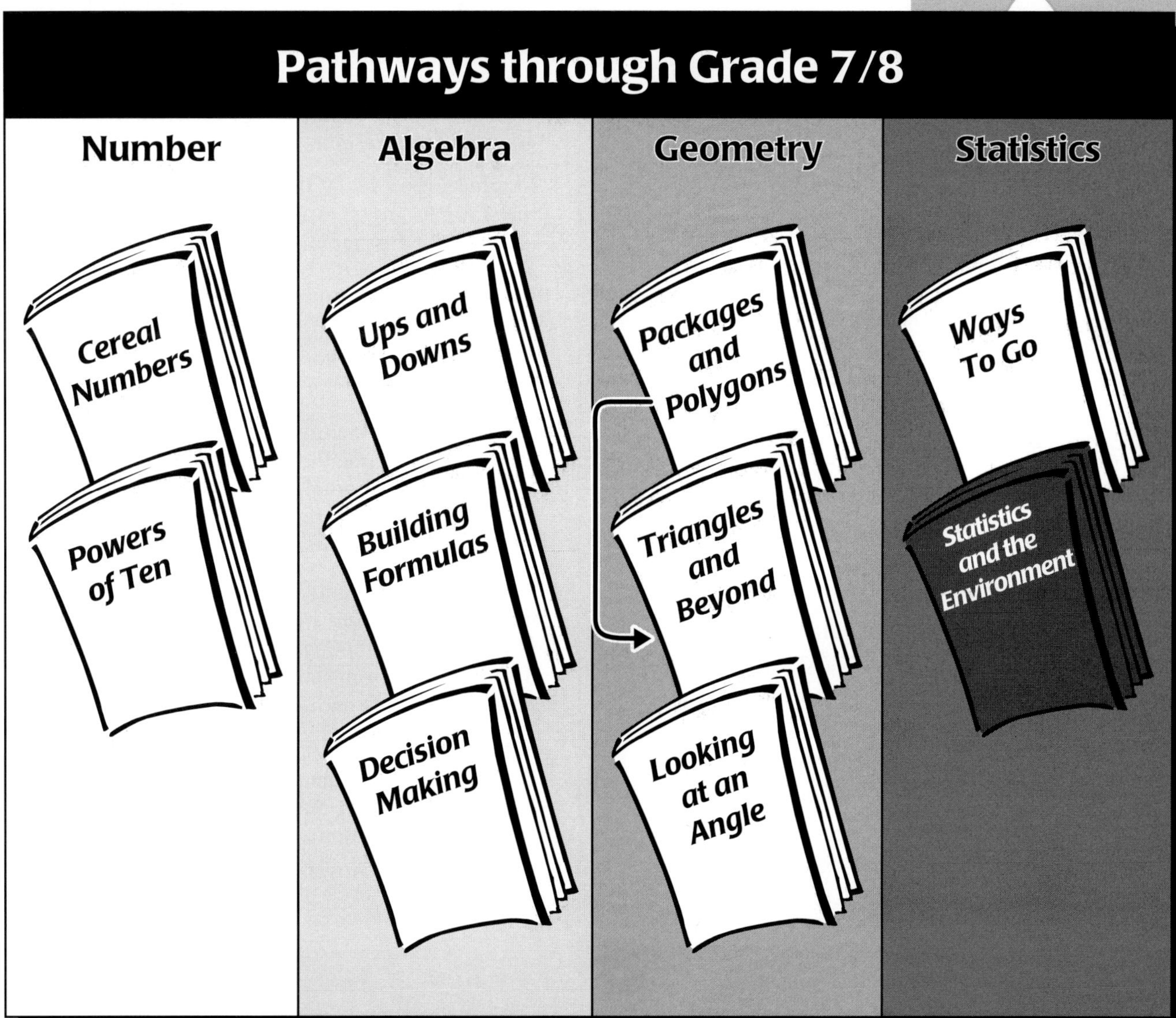

Pathways through the Statistics Strand
(Arrows indicate prerequisite units.)

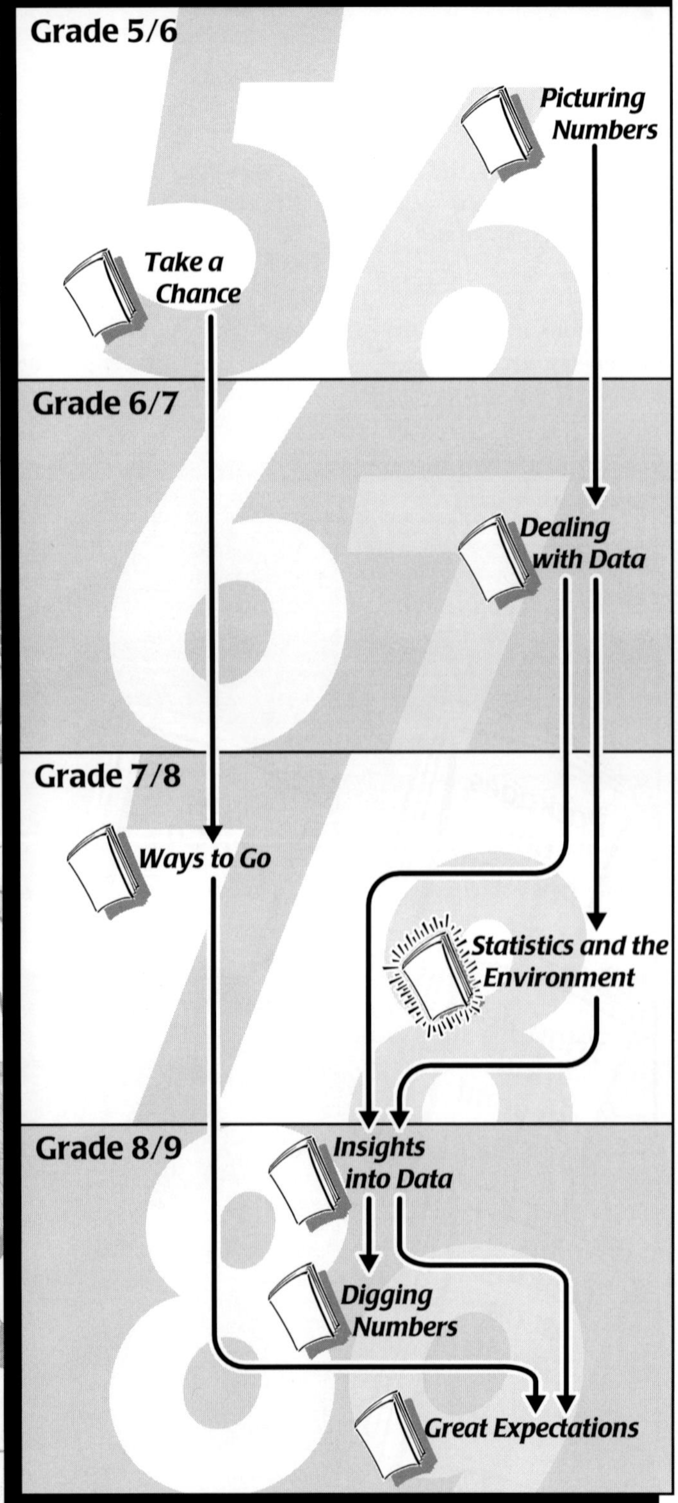

Connections within the Statistics Strand

On the left is a map of the statistics strand; this unit, *Statistics and the Environment,* is highlighted.

The major themes in the statistics strand are dealing with data and understanding chance and probability.

Collecting and displaying data are central to the statistics units. Students should be able to describe data visually and numerically and to draw conclusions based on data and descriptions of data. In *Picturing Numbers,* students are introduced to numerical descriptions of data and to the mean as a measure of center. The mean is then revisited and other summary measures (mode and median) are formally introduced in the grade 6/7 unit *Dealing with Data* and in this unit, *Statistics and the Environment.* These topics are revisited in the grade 8/9 unit *Insights into Data.*

Graphical representations of data are found in all of the statistics units, especially in *Dealing with Data,* in which the histogram, the scatter plot, the stem-and-leaf plot, and the box plot are introduced. Students use data and graphs to draw conclusions and build arguments. Statistics also appear in units from other strands: for example, the mean is used in *Ratios and Rates.*

The critical attitude that is developed in *Picturing Numbers* is extended in the units *Statistics and the Environment* and *Insights into Data.* The focus of *Insights into Data* is the use of summary measures and graphs and making decisions using statistics.

Students were introduced to random samples in *Dealing with Data* and *Statistics and the Environment.* Sampling and drawing conclusions based on data from samples will be revisited in *Insights into Data* and *Great Expectations.*

In the statistics units, a foundation is laid for dealing with data. The probability units *Take a Chance, Ways to Go,* and *Great Expectations* build on students' experiences with data to develop chance and probability concepts.

The Statistics Strand

Grade 5/6

Picturing Numbers
Creating and interpreting different kinds of graphs (bar and line graphs, pie charts, and pictographs), describing data in tabular and graphical forms, and using data as an aid to argument.

Take a Chance
Exploring the meanings of the terms fair and chance, computing chances in situations with few outcomes, using repeated trials to estimate chances, and using tree diagrams to represent two- or three-event situations.

Grade 7/8

Ways to Go
Reading and interpreting different kinds of maps, drawing graphs and networks, representing tables as graphs (and vice versa), solving problems with the help of connectivity graphs and tables, and solving simple combinatoric problems. (*Ways to Go* is primarily a statistics unit but it integrates concepts from the geometry strand, as well.)

Statistics and the Environment
Collecting data; summarizing data with one-number summaries; reading and interpreting different kinds of maps, graphs, and tables; and using graphs and data as aids to argument.

Grade 6/7

Dealing with Data
Creating and interpreting scatter plots, box plots, stem-and-leaf diagrams, histograms, and number line plots; describing data in tabular and graphical forms; using one-number summaries (mean, median, mode, range, quartiles); and using data as an aid to argument.

Grade 8/9

Insights into Data
Analyzing information presented in a graph; gathering data using sampling techniques; recognizing bias in sample surveys; examining misleading representations of data; identifying weak, strong, positive, and negative correlations and no correlation; using lines to summarize the data in a scatter plot; and making and interpreting scatter plots and box plots.

Digging Numbers
Using the properties of height, diameter, and radius to determine whether or not various irregular shapes are similar; predicting length using graphs and formulas; exploring the relationship between three-dimensional shapes and drawings of them; and using length-to-width ratios to classify various objects. (*Digging Numbers* is primarily a statistics unit, but it integrates concepts from the geometry strand, as well.)

Great Expectations
Recognizing the variability in taking random samples and the role of chance in world events; understanding the effect of sample size on the variability in samples; finding the probability of independent and dependent events; using expected values to make decisions; using simulations to investigate probability; and counting all the possible arrangements of a set, with and without repetition.

Connections with Other *Mathematics in Context* Units

Statistics and the Environment is a thematic unit that integrates mathematical concepts from all four strands and from other subject areas, such as science, language arts, and social studies. Students informally investigate the concepts of random sampling and what constitutes a representative sample as they design and conduct surveys and analyze data they have collected along with other data sets in order to design a land-use plan for an environmentally sensitive island. Graphical representations and summary measures of data sets are investigated using techniques developed in *Dealing with Data.*

The following mathematical topics that are included in the unit *Statistics and the Environment* are introduced or further developed in other *Mathematics in Context* units.

Prerequisite Topics

Topic	Unit	Level
bar graphs	*Picturing Numbers*	5/6
	Dealing with Data	6/7
pie charts	*Fraction Times**	6/7
	Picturing Numbers	5/6
scatter plots, histograms, box-plots stem-and-leaf plots, mean, mode, median	*Dealing with Data*	6/7
reading and interpreting graphs	*Picturing Numbers*	5/6
	*Tracking Graphs***	6/7
	Dealing with Data	6/7
metric units	*Measure for Measure* *	5/6
	Reallotment ***	6/7
	Made to Measure ***	6/7
area	*Reallotment* ***	6/7
maps and scales	*Grasping Sizes**	5/6
	Reallotment ***	6/7
absolute and relative comparisons	*Ratios and Rates**	6/7
	*Cereal Numbers**	6/7

Topics Revisited in Other Units

Topic	Unit	Level
reading and interpreting graphs	*Insights into Data*	8/9
	Graphing Equations ***	8/9
exponential growth	*Ups and Downs* **	7/8
	Growth **	8/9
contour lines, profiles, and maps	*Going the Distance****	8/9
tangent	*Looking at an Angle****	7/8
	*Going the Distance****	8/9
statistical measures and graphs	*Insights into Data*	8/9

* These units in the number strand also help students make connections to ideas about statistics.
** These units in the algebra strand also help students make connections to ideas about statistics.
*** These units in the geometry strand also help students make connections to ideas about statistics.

BRITANNICA Mathematics in Context

Student Materials and Teaching Notes

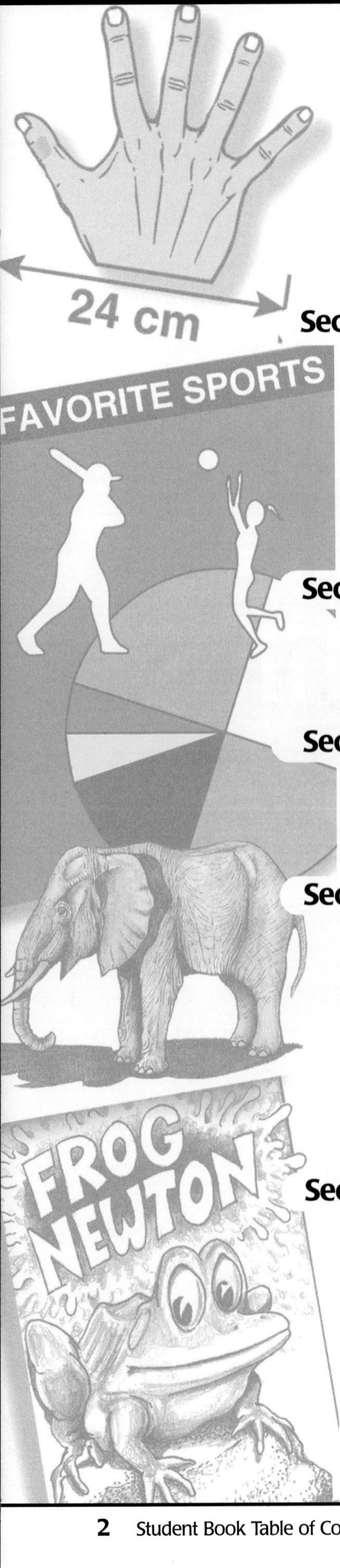

Student Book
Table of Contents

Letter to the Student vi

Section A **Preparing to Use Statistics**
Murre Island 1
Weather 3
Murre Island Bats 5
Topography 6
Describing Murre Island 10
Collecting Data 12
Summary 14

Section B **Data about Endangered Forests**
Murre Island Forests 15
National Forests 18
Summary 20

Section C **Interpreting Data**
Statistical Summaries 21
Vacation Activities 23
Summary 26

Section D **Describing Data**
Environmental Problems 27
Water 28
Energy 31
Being Economical with Electricity 34
Energy Use on Murre Island 34
Waste Disposal 35
Waste Disposal on Murre Island 40
Summary 40

Section E **Land-Use Plans for Murre Island** 41
Campgrounds 44
Hiking Trails 45
Lodge 46
Recreation 47

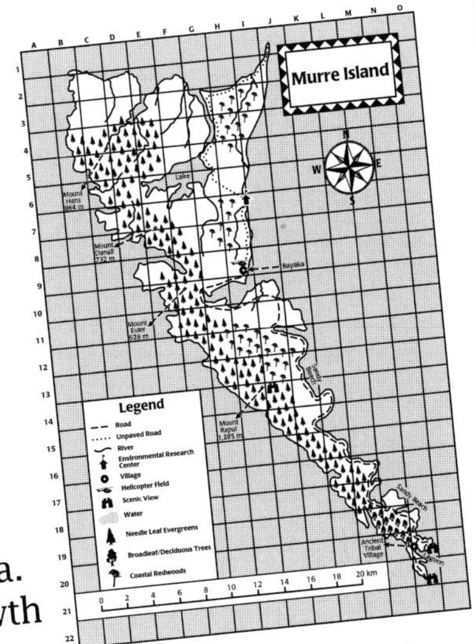

Dear Student,

Welcome to *Statistics and the Environment.*

In this unit, you will develop a plan for turning Murre Island, a fictitious island, into a nature reserve.

To develop an environmentally responsible plan, you will study the island's climate, topography, forests, and wildlife by organizing, displaying, and interpreting data. For example, you will study data on old-growth forests and national parks in order to make decisions about how to preserve old-growth forests on Murre Island.

You will also design and conduct a survey to determine how people like to spend their vacations. You will then tally the results of your surveys and make graphs to display the results. Based on these results, you will decide which facilities on Murre Island will attract visitors.

Finally, you will determine the water, energy, and waste disposal requirements for Murre Island. You will study the water and energy usage and waste production of both your household and of each state in the United States in order to make related decisions about Murre Island.

We hope that you enjoy using statistics to make informed and environmentally responsible decisions.

Sincerely,

The Mathematics in Context Development Team

Work Students Do

The first part of this section introduces two fictional islands and the theme of the unit, which is using statistics to create a plan for a nature preserve. Students read and interpret graphical representations of data and collect their own data. They consider how tourism and recreational activities can be introduced in a way that will have the least impact on the ecology of the island. As in real life, people have differing viewpoints on the conservation of natural resources and on what constitutes ideal land use. Students will form opinions on these issues as they work through this unit, and they are expected to use statistics to support their positions and opinions.

Goals

Students will:

- organize and describe data;*
- use statistical summaries and graphs;*
- calculate with fractions, decimals, and percents;*
- judge whether or not samples are representative and random;*
- read, interpret, and use information from maps or graphical displays of data;
- choose and use appropriate techniques to analyze data;*
- make decisions based on statistical information and justify these decisions.*

* *These goals are introduced in this section and are assessed in other sections of the unit.*

Pacing

- approximately four 45-minute class sessions

Vocabulary

- contour map
- population
- profile
- representative sample
- sample

About the Mathematics

Statistics is the science of collecting, organizing, and interpreting numerical facts that are called *data.* Interest in teaching statistics is due in part to the place that working with data plays in everyday life and in many occupations. For example, farmers study data from field trials of new varieties of crops, and government offices release the latest numerical information on unemployment and inflation. The study of statistics enables students to do computation in meaningful settings. It is also effective in developing quantitative understanding.

Data can be produced by sampling a *part* of a population of people or things in order to gain information about the *whole.* Opinion polls may report the views of the people of the United States based on interviews with a representative sample of about 1,500 people.

Also critical to this section are contour maps and profiles, or vertical cross sections of a landscape (see page 20 of the Teacher Guide). These are two-dimensional representations of a three-dimensional situation. A contour line connects points that have the same height.

Materials

- Letter to the Family, page 120 of the Teacher Guide (one per student)
- Student Activity Sheets 1–4, pages 121–124 of the Teacher Guide (one of each per student)
- graph paper, page 15 of the Teacher Guide (one sheet per student)
- transparency of Student Activity Sheet 3, page 19 of the Teacher Guide, optional (one per class)
- overhead projector, page 19 of the Teacher Guide, optional (one per class)
- colored pencils or markers, page 29 of the Teacher Guide, optional (five different colors per group of students)
- copies of the questionnaire students develop on Student Book page 13, page 31 of the Teacher Guide (5–10 copies per student)

Planning Instruction

You may want to introduce this section with a class discussion about national and state parks and nature preserves. You might ask questions such as *Why do national parks exist? Have any of you visited a national or state park? Were there restrictions on what you could do there?*

Problem 28a may be done as a whole-class activity. Problems 1 and 14 may be done as whole-class activities or in small groups. Students may work on problems 2–6, 10–13, and 22–24 in small groups. Students may work on problems 7–9 and 15–20 individually or in small groups. The remaining problems may be done individually.

Problem 21 is optional. If time is a concern, you may omit this problem or assign it as homework.

Homework

Problems 7–9 (page 12 of the Teacher Guide), 15 and 16 (page 20 of the Teacher Guide), 17 (page 22 of the Teacher Guide), 25–27 (page 28 of the Teacher Guide), 29 (page 32 of the Teacher Guide), and the Extension activity (page 23 of the Teacher Guide) can be assigned as homework.

Planning Assessment

- Problems 18, 19, and 20 can be used to informally assess students' ability to read and interpret information from maps or graphical displays of data.

A. PREPARING TO USE STATISTICS

Murre Island

Imagine that you are a member of a conservation organization called EcoSystem Savers (ESS). You will develop a plan to convert Murre Island, a fictional island, into a nature preserve. Murre Island is modeled after islands in the Pacific Northwest and is named after the murre, a seabird commonly found in the area (and shown in the picture below.)

1. What information or data about Murre Island do you think are necessary in order to develop a plan to convert it into a nature preserve?

Solutions and Samples
of student work

1. Answers will vary. Sample student response:

   To develop a plan you would need information/data about wildlife, climate, and vegetation, as well as maps that give the topography and the size and location of the island.

Hints and Comments

Overview Students are introduced to fictional Murre Island and to the theme of the unit—using statistics to develop a plan for ecological preservation.

Planning Students may read the introduction individually. ESS is an imaginary conservation organization. You may want students to name some real conservation organizations and discuss the goals of these organizations. Problem **1** may be done as a whole-class activity or in small groups.

Comments about the Problems

1. This problem asks students to think about information and data collection. Students may bring up the different aspects of the island that they will investigate in more detail later in this section.

Interdisciplinary Connection You may want students to look for information about the Pacific Northwest in an atlas or encyclopedia.

Did You Know? The murre is a small seabird that has a white body with a black head, neck, back, and wings. These birds nest in large numbers on cliffs along the coasts of the Northern Hemisphere. Murres swim underwater fast enough to prey on small fish. In the fall, the birds swim south.

Source: 1994 Encyclopædia Britannica Micropædia, vol. 8, p. 434.

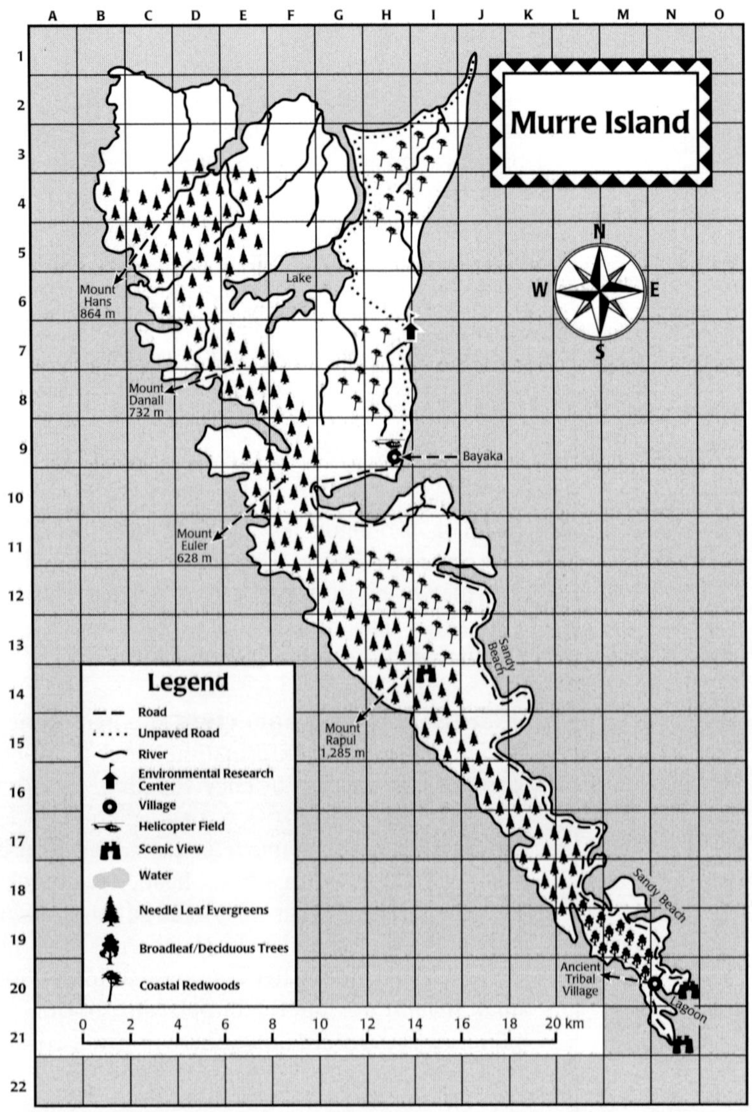

ESS development team members have a lot of difficult decisions to make about Murre Island because it is an environmentally sensitive area. As you can see on **Student Activity Sheet 1,** Murre Island contains a number of old-growth forests. These old-growth forests consist of such ancient trees as coastal redwoods and Douglas firs. The coastal redwood is the world's tallest living thing. Its timber is highly prized, so there is a lot of pressure to cut down these trees.

Orcas and Dall porpoises swim in the seas surrounding Murre Island. The island is also on the migration route of the gray whale. Many species of birds make Murre Island their home.

Although ESS team members want to develop the island solely as a nature preserve, they realize that they also need to accommodate tourists. The money from tourism can be used for environmental protection. Therefore, a land-use plan for Murre Island must also include recreational areas.

2. What are the advantages of including recreational areas in the plan for Murre Island? the disadvantages?

2. Answers will vary. Sample student response:

Advantages:

- Tourists will spend money that can be used for environmental protection.
- Environmental research on the island will get more attention and publicity.
- People will be able to experience forests and wildlife that are endangered.

Disadvantages:

- There will be some environmental damage.
- Recreational areas will mean a loss of natural habitat.
- Loss of natural habitat means a loss of wildlife.
- Tourists will disturb the ancient tribal village.

Materials Student Activity Sheet 1 (one per student)

Overview Students read more about the environment of Murre Island. They describe the advantages and disadvantages of including recreational areas in the plan to develop the island into a nature preserve.

Planning To introduce this page you may ask students who have visited national or state parks or nature preserves what recreational activities they enjoyed there.

Students may work on problem **2** in small groups. You may want students to study the map to locate the types of trees and other items that are mentioned in the map legend.

Comments about the Problems

2. There are some conflicting interests, as is often the case with nature preserves that are open to the public. Perhaps students can relate examples from their own experiences.

Note that students frequently will need the map on Student Activity Sheet 1, not only in this section but also in Section B.

Interdisciplinary Connection You may want students to look for more information on and pictures of the trees and animals mentioned in the text. This study can be guided by the science teacher during science class.

Weather

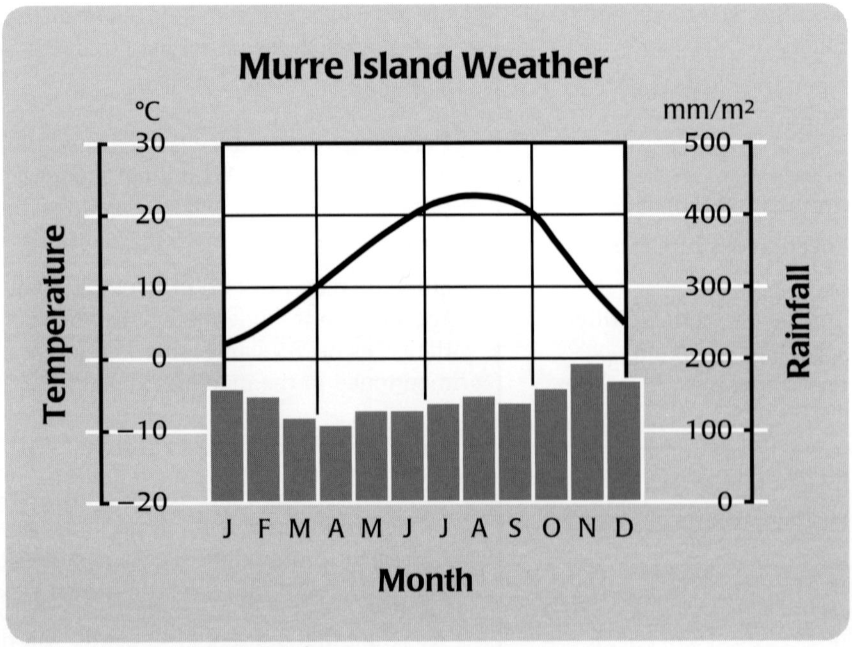

The above graph shows the mean temperature and amount of rainfall for each month of the year on Murre Island. The curve represents temperature in degrees Celsius. The bars indicate rainfall in millimeters per square meter.

3. Do you see a relationship between the temperature and the rainfall on Murre Island? Explain.

4. Estimate the total amount of rainfall (in millimeters per square meter) for Murre Island each year.

5. Write a short description of the weather on Murre Island. Give reasons to support your description.

6. Why might tourists want to visit Murre Island? What time of the year might they want to go? Explain.

Solutions and Samples
of student work

3. Answers will vary. Sample responses:

Yes, there is more rain during the winter months, when the temperature is colder. In the summer, there is less rainfall.

Yes, in the first part of the year, the temperature increases, while the amount of rain decreases. At the end of the year, it is the other way around.

4. Estimates will vary. Accept estimates between 1,700 and 1,800 millimeters per square meter. The amounts per month are about 160 mm + 150 mm + 120 mm + 110 mm + 130 mm + 130 mm + 140 mm + 150 mm + 140 mm + 160 mm + 195 mm + 170 mm, or about 1,755 mm per square meter.

5. Answers will vary. Sample responses:

- The warmest months are June–September, when the mean temperature is 20°C or above.
- There is less rainfall from March to September, except during the month of August.
- It rains at least 100 millimeters per square meter every month.
- The average temperature is never below 0°C.

6. Answers will vary. Sample responses:

Tourists might want to see the ancient trees, the birds, and the porpoises and whales. They may decide to go in April to avoid more rain.

Some people like to hike. Good temperatures for a hike are between 15°C and 20°C. So people who want to hike might want to go in May, June, or October. In these months it is not raining very much either.

Hints and Comments

Overview Students read and interpret graphs showing the mean monthly temperature and the total amount of rainfall on Murre Island.

About the Mathematics The information about the mean temperature is given by a line graph. Students first investigated line graphs in the unit *Tracking Graphs*. They should be able to interpret this graph globally. They do not need to know how these mean temperatures were calculated; however, they must understand that the mean temperature of one day can be equal to, more than, or less than the mean temperature of the month in which it occurs.

The bar graph shows the total amount of rainfall for each month, measured in millimeters per square meter. This total may be the mean of totals for that month during different years, or it may represent only one year. Students worked with bar graphs in the units *Picturing Numbers* and *Dealing with Data.*

Planning You might want to start with a class discussion about the graph. Ask students what information they can find in the graph and how this information is presented. You may have students compare Murre Island's climate with the climate in their area. Students may work in small groups on problems **3–6.**

Comments about the Problems

3. Note that the rainfall is measured in millimeters per square meter and the temperatures in degrees Celsius.

5. Students should translate the graphical data into a written description. Make sure they use numbers to quantify their words.

6. Students have only weather data with which to answer this question. They will use information from the maps later in the unit.

ESS is also creating a nature refuge on another island, called Albatross Island.

Albatross Island is rocky and windswept. It is covered in tussock grass that can grow up to 9 feet high. The island is home to a large breeding colony of black-browed albatross, as well as a number of rockhopper penguins.

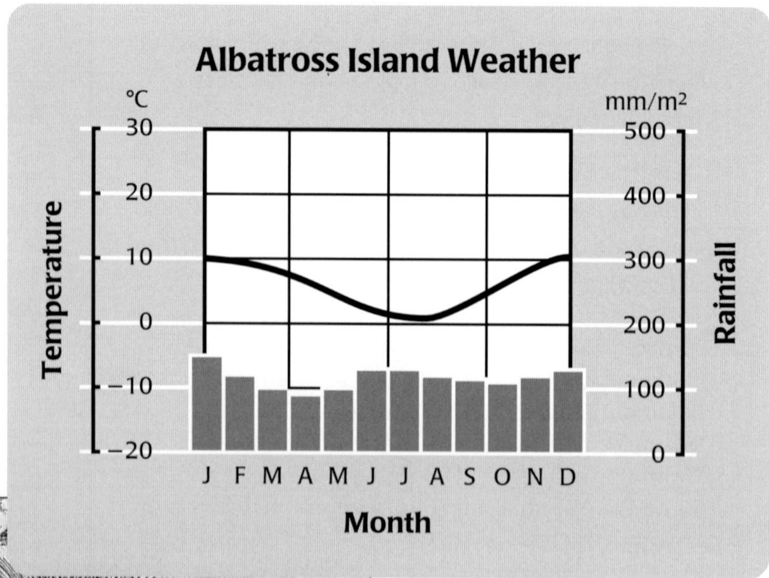

The above graph shows the mean temperature and amount of rainfall for each month of the year on Albatross Island. The curve represents temperature in degrees Celsius. The bars indicate rainfall in millimeters per square meter.

7. Compare the weather on Albatross Island with the weather on Murre Island.

8. In which hemisphere, the Northern or Southern, might Albatross Island be located? Explain.

9. Which island do you think would attract more tourists as a nature refuge, Murre Island or Albatross Island? Explain.

7. Answers will vary. Sample response:

   Albatross Island is fairly cool. The mean temperature varies from about 0°C to 10°C, while on Murre Island, the mean temperature varies from about 2°C to 23°C. In general, Murre Island is warmer and has more variation in temperature. When it is warm on Murre Island, it is cold on Albatross Island and vice versa. The precipitation on Albatross Island is relatively constant year-round; still, it is not as rainy as Murre Island.

8. In the Southern Hemisphere. Students should note that the colder, winter temperatures are in June, July, and August. Some students may also know that penguins live only in the Southern Hemisphere.

9. Answers will vary. Sample response:

   Murre Island is more attractive because the temperatures are higher and there is more variation in wildlife and vegetation. The giant trees on Murre Island are more awe-inspiring than Albatross Island's tall grass.

Overview Students investigate temperature and rainfall graphs of another island. They compare the weather data on the two islands and draw conclusions about the location of the second island.

Planning Students may work on problems **7–9** individually or in small groups. Problems **7–9** may be assigned as homework. Discuss students' answers for all three problems.

Comments about the Problems

7–9. Homework These problems may be assigned as homework.

7. "Summer" is a relative concept when connected with months. You can discuss whether it is possible to say things such as "on Murre Island it is colder in summer than in winter."

Interdisciplinary Connection You may want students to look for information about tussock grass, black-browed albatross, and rockhopper penguins during science class.

Murre Island Bats

ESS is developing a variety of nature walks on Murre Island so that visitors can see some of the island's rare animals.

One fascinating animal that can be found only on Murre Island is the Rapul bat, pictured above. Many of these long-eared, pale-gray bats live in the caves on Mount Rapul. After dark, they can be seen picking insects from bark and leaves.

Scientists from the Environmental Research Center on Murre Island think that the temperature determines the length of time that the bats stay out of their caves. The scientists observe the bats for one week in August during the evening and record the temperatures and lengths of time that the bats spend outside the caves.

Bat Statistics (August 1–7)

Date	Temperature (in °C)	Time outside Caves (in minutes)
August 1	20	16
August 2	22	21
August 3	19	15
August 4	22	24
August 5	26	30
August 6	23	23
August 7	19	14

Use the above table to answer the following questions:

10. Is there a relationship between the temperature and the length of time that the bats stay outside the caves? Explain.

11. **a.** Make a scatter plot of the data in the above table with temperature on the horizontal axis and time on the vertical axis.

b. Describe the relationship represented by the points in the scatter plot.

12. Based on the scatter plot you made in problem **11a,** how many minutes might the bats fly outside the caves on an evening when the temperature is 21°C? Explain how you found your answer.

Solutions and Samples
of student work

10. Yes, there appears to be a relationship. It seems that the higher the temperature, the longer the bats fly.

11. a.

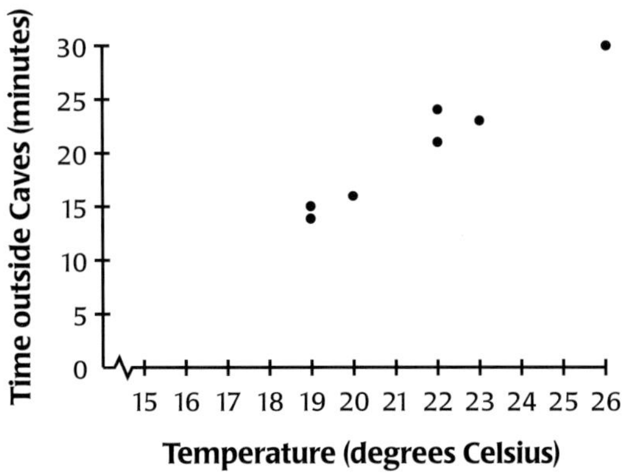

b. Descriptions will vary. Some students may say that the points appear to follow a diagonal line.

12. Answers may vary. A good prediction based on the arrangement of the data points in the scatter plot would be about 19 minutes.

Hints and Comments

Materials graph paper (one sheet per student)

Overview Students investigate the behavior of bats in relation to evening temperatures. They make and analyze a scatter plot of the data.

About the Mathematics A useful graph for representing qualitative variables is a scatter plot. A scatter plot is a plot of observations as points in the plane. To interpret a scatter plot, look first for an overall pattern. This pattern should reveal the direction, form, and strength of the relationship between the variables. Students might find a line by placing a transparent ruler over the scatter plot and moving it until the edge seems to give a line that summarizes the data. One criterion for a good summary line is to have about half the points on either side of it. Students studied scatter plots in the unit *Dealing with Data.*

Planning Students may work in small groups on problems **10–12.** Discuss students' answers to problems **10** and **11** before they continue with problem **12.**

Comments about the Problems

11. b. The relationship seems to be linear, but do not make an issue of drawing a line of best fit at this point. Students should see that the higher the temperature, the longer the bats stay out.

12. You might discuss the difference between using the table and using the scatter plot to make the prediction. With the table, it is difficult to see a pattern because the points are not in any order. In the scatter plot, you see a gap around 21°C. The flight time could be any time between 16 and 21 minutes.

Did You Know? Bats are able to find their way about at night with great ease. This ability has always amazed people and may help to explain some of the myths and legends that have grown up around bats. While in flight, a bat will emit a series of extremely high-pitched squeaks or clicks. One reason that bats use high-frequency sounds is that they produce coherent echoes from small objects, such as flying insects, that enable bats to pinpoint their locations. Most echolocation sounds of bats are inaudible to humans. You can read more about bats in Laurence Pringle's *Batman: Exploring the World of Bats* (New York: Scribner, 1991).

TOPOGRAPHY

ESS is interested in setting up hiking trails in the mountain range that runs along the west side of Murre Island and needs to know both the elevation and the steepness of the mountains. This information can be found on the *contour map* of Murre Island on **Student Activity Sheet 2,** which shows the topography of the island.

Use the Murre Island map on **Student Activity Sheet 1** and the contour map on **Student Activity Sheet 2** to answer the following questions:

13. a. What is the highest mountain on Murre Island?

b. Where on the island is the highest mountain located?

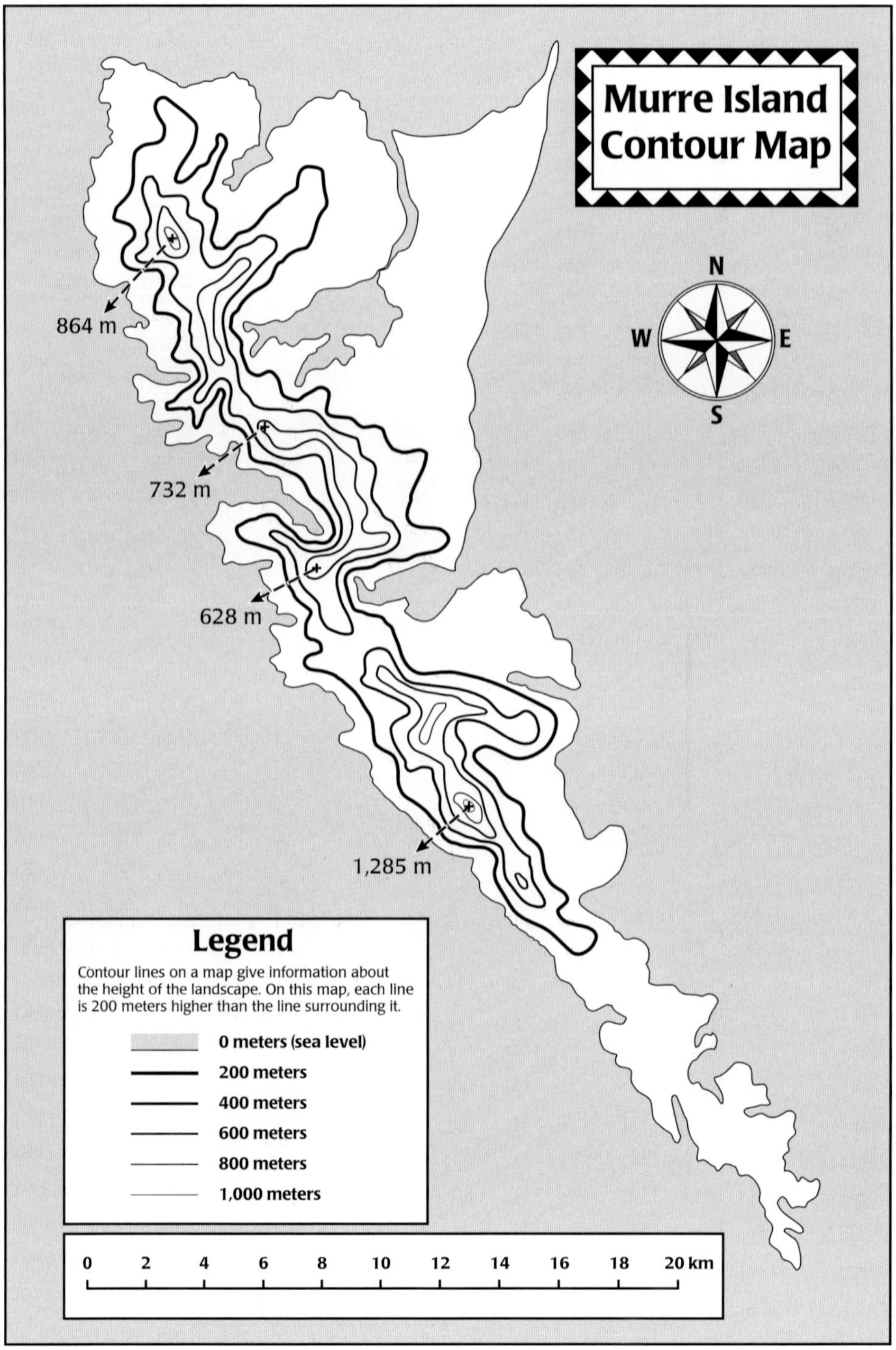

Solutions and Samples
of student work

13. a. Mount Rapul (1,285 meters)

b. Descriptions may vary. Sample answers:

- The highest mountain is located about $\frac{1}{3}$ of the way from the southeastern end of the island to the northwestern end.
- The highest mountain is located about $8\frac{1}{2}$ kilometers southeast of Bayaka.

Hints and Comments

Materials Student Activity Sheet 2 (one per student)

Overview Students investigate a contour map of Murre Island.

Planning You may want to discuss the features of the contour map with your students. Talk about the information each line gives. Students may have had some experiences with contour lines on a map in social studies class. Students may work on problem **13** in small groups.

Comments about the Problems

13. This question is to be answered by reading and interpreting the data on the maps. Encourage students to look up the word *topography* in a dictionary if it is an unfamiliar word.

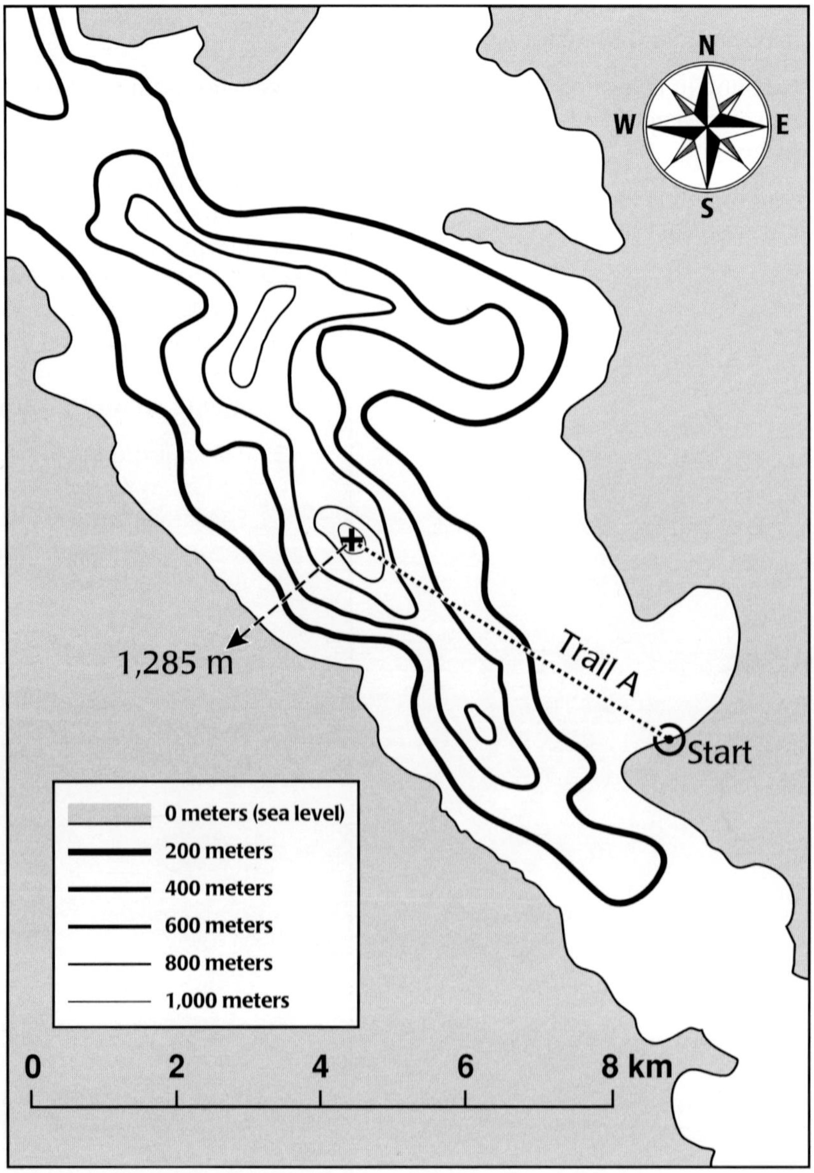

Pictured above is an enlarged portion of the contour map from **Student Activity Sheet 2.** It shows one of the proposed hiking trails.

14. a. Use the scale to find the length of the hiking trail.

b. How many total meters does the hiking trail ascend?

c. A side view of the trail up the mountain is called a *profile.* On **Student Activity Sheet 3,** draw a profile of trail A from the start to the summit.

d. Which parts of the trail are difficult to climb? Why?

Solutions and Samples
of student work

14. a. about 5 kilometers

b. 1,285 meters (You start at sea level and go all the way up.)

c.

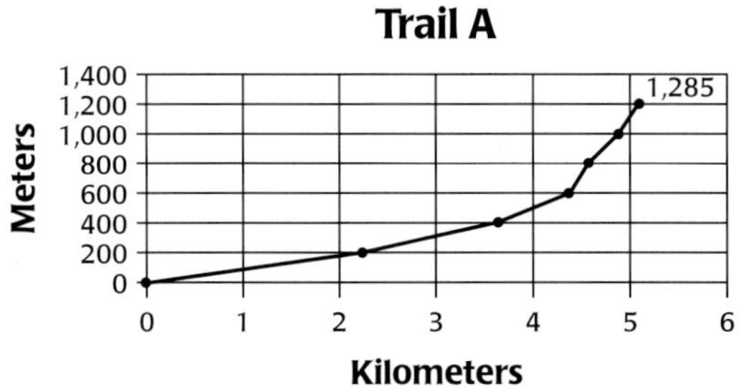

d. The trail gets harder to climb toward the top. In the final kilometer of the trail, it is much steeper.

Hints and Comments

Materials Student Activity Sheet 3 (one per student); transparency of Student Activity Sheet 3, optional (one per class); overhead projector, optional (one per class)

Overview Students use the scale on the contour map to find the length of a trail. Then they draw a profile of the trail.

About the Mathematics A profile is a vertical cross section of the landscape. A sophisticated way to draw a profile is by using the scale on the map for the horizontal scale of the profile and placing the edge of a piece of paper along the trail to mark the points of changes in elevation. The paper can then be placed along the horizontal scale of the profile and elevations can be marked perpendicular to the points on the paper. Contour lines and profiles are revisited in the unit *Going the Distance.*

Planning It may be necessary for students to talk about the contour map, verbally describe the trail, and sketch the profile before drawing the profile in more detail. A copy of Student Activity Sheet 3 on a transparency may be useful for such discussions. Problem **14** may be done as a class or in small groups.

Comments about the Problems

14. a. Students can use a paper strip to mark the length of the trail on the map and then use the scale line to find the trail's actual length.

c. Some students may have difficulty drawing the profile. Let them struggle for a short while before you discuss this in class.

d. Students may generalize that the steepest part of the trail can be found where the contour lines are closest. This will be discussed more explicitly in problem **16b** on the next page.

Interdisciplinary Connection If this unit is being taught as an interdisciplinary unit, problem **14** and problems **15–17** on the next pages can be completed in a social studies class.

Three suggested trails through the mountain range are shown on the map below.

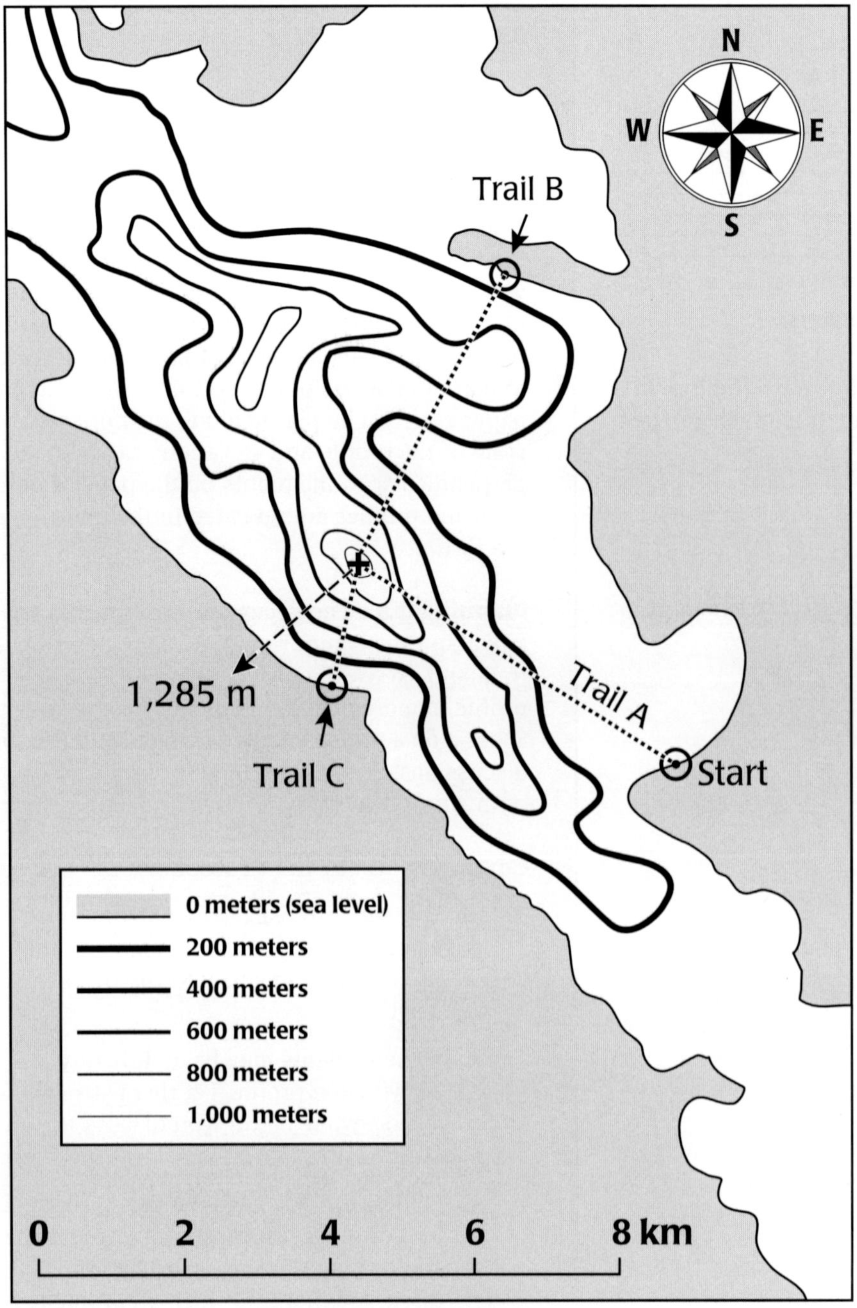

15. a. How do the trails differ?

b. Which trail would you rather hike? Why?

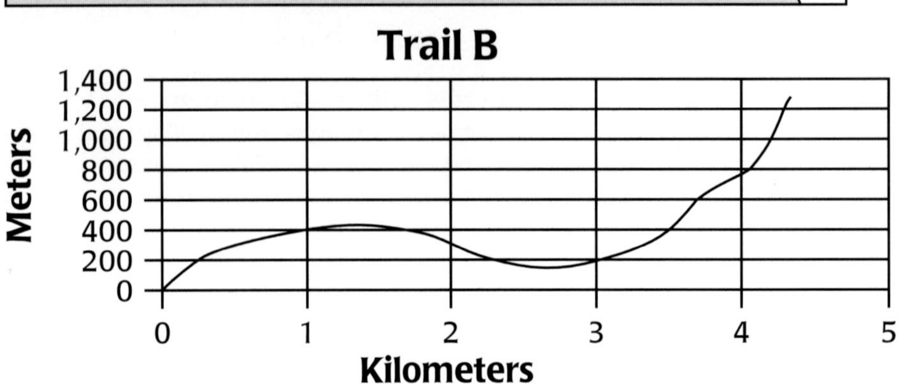

16. a. Indicate the steepest part of trail B on both the profile and the contour map on **Student Activity Sheet 4.**

b. How can you determine the steepest part of a trail using only the contour map?

Solutions and Samples
of student work

15. a. Students' answers may vary. Some students may give a more global comparison: The trails differ in length and in steepness. Some students may describe these differences in more detail. They may notice that trails A and C are both ascending all the time, while trail B is first going up, then going down, and finally going up until it reaches the summit.

b. Answers will vary. Some students may choose trail B because you can take a break from climbing and descend for awhile in the middle of the trail. Others may choose trail C because they want to get to the summit faster, and this trail is short and steep, while still others will choose trail A because this trail gradually becomes steeper.

16. a. Students should indicate the last part of this trail as the steepest part. See the map and profile below.

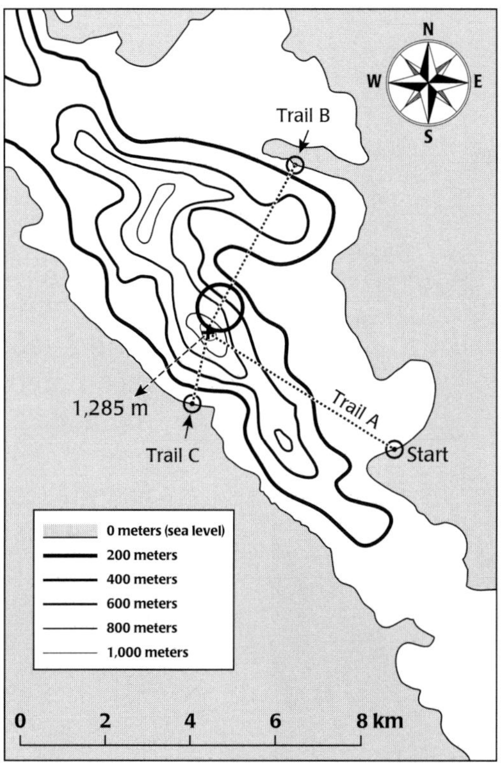

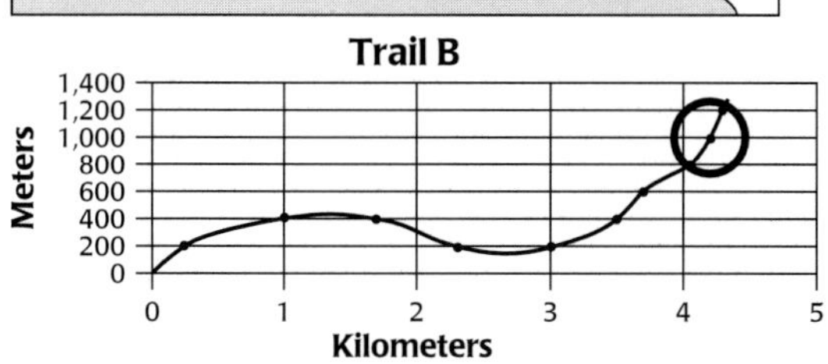

b. The contour lines on either side of the steepest part are closest to each other.

Hints and Comments

Materials Student Activity Sheet 4 (one per student)

Overview Students compare three trails drawn on the contour map. Then they use the contour map and the given profile of trail B to find the steepest part of that trail.

About the Mathematics The steepness of any part of a trail is determined by both the vertical and the horizontal distances covered. The ratio between the vertical and horizontal distances is the slope, or the tangent of the slope angle. The path with the greatest slope is the steepest. Students learn about the concept of tangent in the unit *Looking at an Angle,* and this concept is revisited in the units *Graphing Equations* and *Going the Distance.*

Planning Students may work on problems **15** and **16** individually or in small groups. Discuss students' solutions and strategies for both problems. Problems **15** and **16** may be assigned as homework.

Comments about the Problems

15–16. Homework These problems may be assigned as homework. These questions are answered by reading and interpreting graphical representations of data.

15. At this point in the section, most students should be able to interpret the contour lines on the map. If some students are still having difficulties, have them write the height numbers next to the contour lines on the map on Student Activity Sheet 4.

16. b. Be sure students understand that the steepest part can be found just by looking at the contour lines. This can be explained as follows. It is about 2 kilometers on trail A from the start to the first contour line, which is 200 meters above sea level. Thus, on this trail you would climb 200 meters in the first 2 kilometers, as opposed to on trail B, where you would climb 200 meters in the first $\frac{1}{2}$ kilometer.

The forested mountain slopes on Murre Island offer many spectacular views of the beautiful Pacific coastal wilderness area.

17. a. Indicate the part of trail B that descends on both the profile and the contour map on **Student Activity Sheet 4.**

b. How can you determine the part of a trail that descends using only the contour map?

18. Draw the profile of trail C on **Student Activity Sheet 5.**

19. Compare the profiles of trails A, B, and C. Which of the three trails would you recommend for the nature preserve on Murre Island? Explain your decision.

17. a. See the contour and profile maps below.

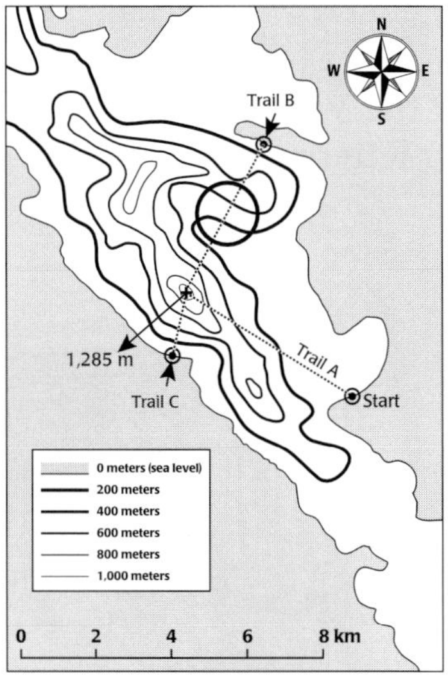

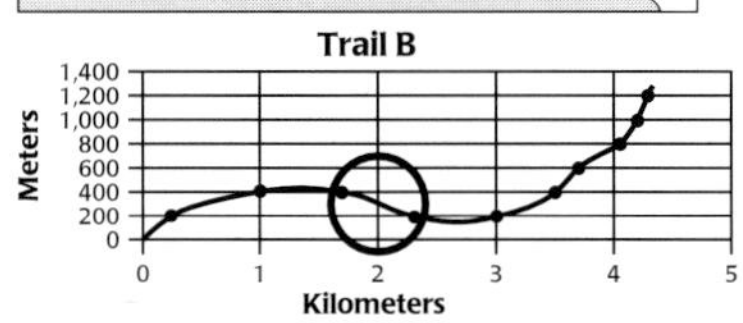

b. If the trail moves from a contour line that represents a higher elevation to a contour line that represents a lower elevation, then it is descending.

18.

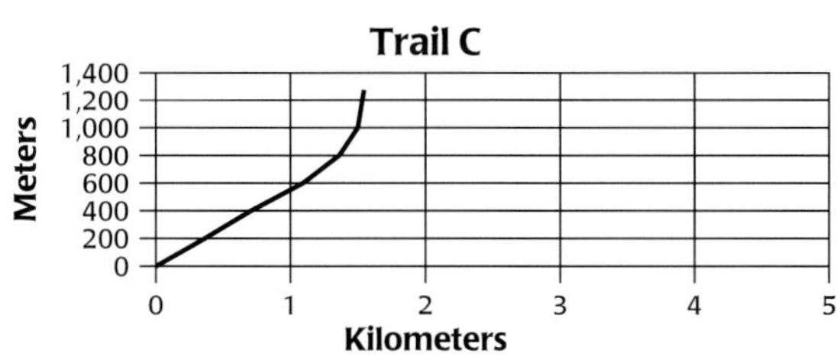

19. Answers will vary. Students may use one or more of the following arguments.

Trails A and B would be less strenuous for visitors. These trails have gradual ascents and start near a road. Trail A travels through needle leaf evergreen forests; trail B travels through coastal redwood and needle leaf evergreen forests and may offer more variation to the visitors. Trail C has a steep ascent, and the start is not near a road.

Trails A, B, and C can be rated on a scale from easy to difficult. (Trail A might be rated easy to moderate because it has a gradual ascent from 0 to 600 meters, while the last section is very steep. Trail B might be rated moderate to difficult because although the first $2\frac{1}{2}$ kilometers are less strenuous, the final $1\frac{1}{3}$ kilometers are very steep. Trail C is very difficult because it ascends steeply from 0 to 1,285 meters in $1\frac{1}{2}$ kilometers.)

Materials Student Activity Sheets 4 and 5 (one of each per student)

Overview Students identify the descending part of trail B. They draw the profile of trail C. Finally, they use the profiles of the three trails to decide which of the trails to recommend for the nature preserve.

Planning Students may work on problems **17–19** individually or in small groups. Problem **17** may be assigned as homework.

Comments about the Problems

17. Homework This problem may be assigned as homework.

17–19. These questions are answered by reading and interpreting the contour maps and profiles.

18–19. Informal Assessment These problems assess students' ability to read and interpret information from maps and graphical displays of data.

Extension Give students the profile of a trail that ascends to the summit of Mount Rapul in a straight line and descends in a straight line to a point different from the starting point. Ask students to draw the trail on a contour map of Murre Island from one of the student activity sheets.

It may be easier to find the trail if the scale of the horizontal axis of the profile is the same as the scale on the map. Then the students can position the profile on the map with the highest point on the summit. The ascending and descending parts of the trail can be found by rotating the paper around this point until the kilometer marks on the horizontal axis of the profile match up to the contour lines.

Describing Murre Island

The board members of ESS want to raise money for the Murre Island conservation project. They plan to send fundraising letters describing the island to a variety of organizations.

EcoSystem Savers

20. Using the information from this section, write a description of Murre Island to include in the letters. Be sure to answer the following questions in your description:

- How big is the island?
- Is it possible to walk to all of the main parts of the island? What other kinds of transportation may be necessary?
- How high is the highest point on the island? How high are the other summits on the island?
- How big is the lake?
- How many kilometers of road are on the island? How much of the road is paved?
- Which part of the island is covered with forest?
- How many accessible beaches are there on the island?
- Which recreational activities are possible on the island? Where?

To make your description even more interesting, you can include drawings.

Solutions and Samples
of student work

20. Descriptions will vary. Sample student responses appear under each category listed below.

Size

- The length of Murre Island is 43–44 kilometers.
- The width the island varies from about $2\frac{1}{2}$ kilometers to $14\frac{1}{4}$ kilometers.
- The island's total area is about 272 square kilometers. This is based on counting 68 squares, each of which is 4 square kilometers. Since students may use different counting techniques, a range of 65–70 squares is reasonable.

Transportation

- No, you cannot climb all of the mountains or walk along the southwestern shore.
- Students may suggest horseback riding, all-terrain vehicles, kayaks, or mountain bikes.

Mountains

- Mount Rapul: 1,285 meters
- Mount Hans: 864 meters
- Mount Danall: 732 meters
- Mount Euler: 628 meters

Lake

- The lake covers about $2\frac{1}{2}$ square kilometers

Roads

- Murre Island has about 65 kilometers of road; 43 kilometers of paved road, 22 kilometers of unpaved road.

Forests

- The southwest coast and the mountains along the center of the island are covered with forests, as is the major peninsula jutting north.

Beaches

- There are two accessible sandy beaches on the island

Recreational Activities

- hiking, kayaking, wildlife viewing, and swimming.

Hints and Comments

Overview Students read and interpret the maps to describe specific aspects of Murre Island.

Planning Students may work on problem **20** in small groups or individually. This problem can also be used as informal assessment.

Comments about the Problems

20. Informal Assessment This problem assesses students' ability to read and interpret information from maps or graphical displays of data.

Writing the description will require that students calculate length, width, and area using the scale.

Interdisciplinary Connection If this unit is being taught as an interdisciplinary unit, problems **20** and **21** (on the following page) can be completed in a language arts class.

After sending out the fundraising letter, ESS receives the following reply:

FOSSIL FUELS INC.

Sandy Williams, Director
ESS

Dear Ms. Williams:

We read about your Murre Island conservation project with interest. As you may know, Fossil Fuels' reputation has been damaged as a result of the oil spill from one of our super-tankers that occurred two years ago. Although we have tried to compensate the victims of this disaster as much as possible, we want to start a new promotional campaign to boost the image of our company.

We would like to launch our campaign by supporting your Murre Island project, and we are willing to fund part of your work. The idea of using tourism as a financial source for protection of the environment sounds very promising. However, we wish to cooperate with your organization on this project only under certain conditions. For example, we want a hotel named after our company built on the island.

For the development of the project, we suggest that you contact Quick Investment Real Estate Company. We have had good experiences with this company in the past.

We look forward to meeting you to discuss a more detailed proposal that we can present to our board of directors.

Sincerely,

Oliver Sax, President
Fossil Fuels, Inc.

21. Write a letter on behalf of ESS in response to Mr. Sax. Remember that ESS needs financial help in order to develop Murre Island into a nature preserve. Consider whether you should agree to Mr. Sax's conditions and whether a real estate company should plan the development on the island.

You may want to do some research to determine whether there are state and federal land development laws that must be considered.

Solutions and Samples
of student work

21. Letters will vary. Sample student response:

Dear Mr. Sax:

We at ESS are happy to hear that you want to help with our conservation efforts on Murre Island. Your involvement will definitely help improve the image of Fossil Fuels. We will be happy to name a hotel the Fossil Fuels Inn, or whatever you wish, as long as you are paying for the hotel's construction and the profits go to protect the island environment. The hotel's design and placement will be determined by a study of the ecology of the island. The development must also follow strict federal regulations regarding construction on designated wildlife preserves. When we have a proposal ready, I'll telephone you to set up a meeting with your board of directors.

Sincerely,

Sandy Williams

Hints and Comments

Overview Students write a letter to the president of an oil company as part of a discussion on how the island can be developed with private funds.

Planning Problem **21** is optional. If time is a concern, you may omit this problem or assign it as homework.

Comments about the Problems

21. Students may want to research state and federal laws and regulations that would impact land-use decisions. Zoning laws and local ordinances must be obeyed. Private landowners and tribal councils may need to be included in the decision-making process.

You may choose not to have students write a reply to Mr. Sax, but to discuss the different factors that need to be balanced in the development of Murre Island.

Collecting Data

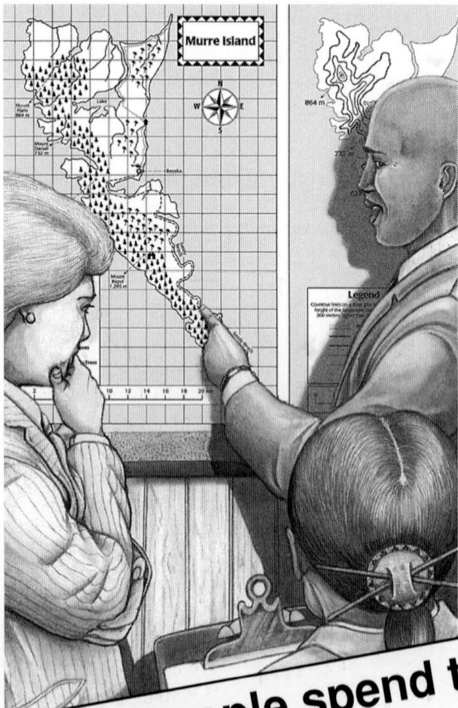

In order to determine what kinds of housing and recreational facilities to build for tourists on Murre Island, ESS decides to find out how people spend their vacations. The organization wants to know such things as the preferred time of year to vacation, the preferred place to stay, and preferred recreational activities.

22. How can ESS gather the needed information?

Shown below is a pie chart that was printed in the newspaper *USA Today.*

23. What percent of the people surveyed spend their vacations away from home?

24. a. Survey your class to find out how students spend their vacations.

b. Make a pie chart of your results similar to the one on the left.

c. Are the results from your class the same as the results from the newspaper survey? Explain.

USA Today surveyed 686 people to get the information for its pie chart. (Notice the line at the bottom of the graph which states, "Representative sample of 686 adults.")

25. Would "How People in the United States Spend Their Vacations" be a better title for the pie chart? Explain.

26. Suppose that ESS wants to display data in pie charts with the following titles. Who should be interviewed and why?

a. How Middle School Students Spend Their Vacations

b. How People in The Netherlands Spend Their Vacations

c. How Adults Spend Their Vacations

27. Could ESS use the pie chart from *USA Today* for any of the above surveys? Explain.

Solutions and Samples
of student work

22. Answers will vary. Students may suggest that they use a questionnaire or do some other kind of survey. They might also call a travel agency.

23. 73%. Strategies will vary. Sample strategies:

- 100% – 21% – 6% = 73%
- 30% + 17% + 26% = 73%

24. a. Survey results will vary, but students' survey categories should be identical to those on the pie chart.

b. Pie charts will vary, depending on the results of the class survey.

c. Answers will vary. Most likely, the class's survey results will differ from the newspaper's survey results. The class is a different *population* and a nonrepresentative sample of the United States.

25. Answers may vary. It is not clear whether the sample included only people from the U.S.; however, the sample is described as "representative," and the pie chart is from *USA Today,* which suggests that it is representative of the entire United States.

26. Answers may vary. Sample student responses:

a. students in different middle school grades from different parts of a state and from different states and countries (if the countries have middle schools)

b. a random selection of the whole population of The Netherlands

c. a random selection of adults worldwide

27. ESS could use the *USA Today* pie chart for How Adults Spend Their Vacations, but a more Murre Island–oriented questionnaire could give more valuable information.

Hints and Comments

Materials colored pencils or markers, optional (five different colors per group of students)

Overview Students investigate a pie chart that shows how people spend their vacations. The students then survey their class about the same topic and compare their results with those shown in the given pie chart. Students also explain who should be interviewed in order to answer certain questions.

About the Mathematics
In the unit *Fraction Times,* students make a pie chart by using a bar. They color parts of the bar to represent each group of people or items. Then they curl the bar into a circle. In this way, the different portions of the circle can be easily marked.

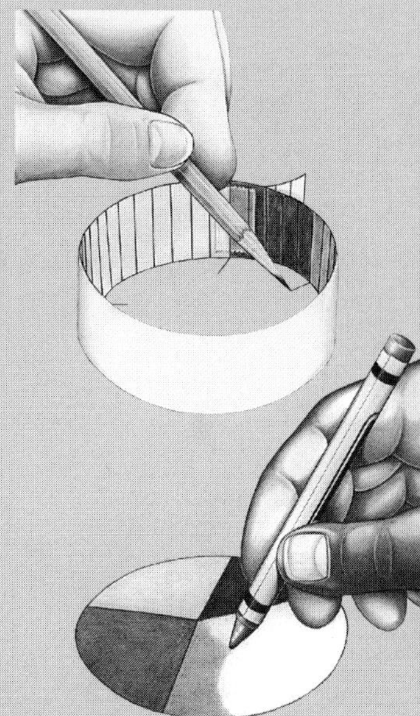

Planning
You can introduce this lesson with a discussion about what ESS would want to know before developing more detailed plans for tourism on Murre Island. Then have students work on problem **22–24** in small groups. Problems **25–27** can be completed individually or assigned as homework.

Comments about the Problems

22. If students say "ask people," try to get them to think about which people and how many. Introduce the word *sample* here. Hint at the idea of taking a random sample.

25–27. Homework These problems may be assigned as homework.

25. You can discuss why samples are used instead of asking the entire population. Cost, time, and difficulty are some possible reasons.

26. Discuss what must be taken into account when choosing samples in each of these cases.

ESS decides to conduct a detailed survey to find out how people spend their vacations. The survey team wants to collect data to answer the following questions:

- What do people like to do while on vacation?
- Do people of different ages like to do different things on vacation?
- Do people like to stay at hotels, campgrounds, or lodges?
- With whom do people prefer to travel: tour groups, family members, or friends?

28. a. Create a survey for ESS. Consider the data that the organization wants to gather. Then determine a format that will make each question easy to answer and the answers easy to tally and interpret.

b. Over the next week, ask 5 to 10 people to answer your survey. Select a variety of respondents so that your results provide a clear picture of how people spend their vacations.

Note: You will need your completed surveys again in Section C.

28. Surveys will vary. Sample survey :

Sample Questionnaire

We want to know something about how people spend their vacations. Please answer these five questions.

1. What is your age?

   ____

2. Check one:
   - ❑ male
   - ❑ female

3. With whom do you travel?
   - ❑ family
   - ❑ tour group
   - ❑ friends

4. Where do you usually spend the night?
   - ❑ in a campground
   - ❑ in a hotel, motel, or lodge
   - ❑ on a boat
   - ❑ in a cabin
   - ❑ other

5. Which of the following activities do you like to do during your vacation? You may choose more than one.
   - ❑ swimming
   - ❑ bicycling
   - ❑ waterskiing
   - ❑ snow skiing
   - ❑ mountain climbing
   - ❑ sailing/canoeing
   - ❑ horseback riding
   - ❑ hiking
   - ❑ hunting/fishing
   - ❑ other

Materials copies of the questionnaire students create for problem **28a** (5–10 copies per student)

Overview Students create a questionnaire and carry out a survey.

Planning You can use the sample survey (see the solutions column) or you may ask students to make up their own survey. All students should use the same questionnaire. Be sure that the questions that are listed below are included in their survey. Set a deadline for bringing the results of the survey to class. Complete problem **28a** as a class. Students should work on part **b** individually.

Comments about the Problems

28. The questions listed below are important for the analysis of the data in Section C.

- "What is your age?" This is an open-ended question. An answer blank is appropriate. The specific ages of people in the survey are important in Section C.
- "With whom do you travel on vacation?" This can be open-ended or have a multiple-choice format. The multiple-choice format allows students to structure the categories they want to use in the analysis of the data and provides for more efficient tallying of survey results. In order for the analysis of survey results in Section C to be effective, it is important to use the following categories: with family, with friends, and with tour groups.
- "Where do you stay overnight?" This can be open-ended or have a multiple-choice format. The categories used in the analysis in Section C include hotel/motel/lodge, campground, cabin, boat, and with relatives.
- "What activities do you do on vacation?" This can be open-ended or in a checklist format. This part of the survey should include a statement that the person should check all activities they do on vacation, not those that they would like to do on vacation. It is also appropriate to put a blank for other activities the person wants to list. A checklist format is helpful in tallying the survey results for this question.

Summary

As a member of ESS, you have made many important decisions about how to best create a nature preserve on Murre Island. You've used a number of statistical techniques and tools, including bar graphs, line graphs, scatter plots, and surveys.

Summary Questions

29. Describe the statistical techniques that you used to study each of the following:

a. the time of year that people are most likely to visit Murre Island

b. the lengths of time and the temperatures at which Rapul bats are out of their caves

c. the steepness of different hiking trails

d. the different kinds of facilities that tourists prefer

29. a. Students used the mean temperatures of the months as shown in a line graph and the average rainfall for each month as shown in a bar graph. They determined when the ideal temperature and the least amount of rain might occur.

b. Students plotted data from a table to create a scatter plot. With this scatter plot, it was possible to interpolate data missing from the table.

c. Students used a contour map to make profiles of three given trails. By comparing the profiles, students were able to determine which trail is the steepest.

d. Students conducted a survey to determine whether people like to stay in hotels, campgrounds, or lodges.

Overview Students read and discuss the Summary, which reviews the statistical techniques and tools used in this section. They then describe specific uses of these tools.

Planning Have students read and discuss the Summary. Problem **29** may be completed individually or assigned as homework.

Comments about the Problems

29. Homework This problem may be assigned as homework. You may want to discuss how the different representations can be used in relation to the developing plan of Murre Island. It is important that students describe how they used the different representations.

Work Students Do

In this section, students interpret a map of Murre Island to determine the types of trees found there and the area of the island that is covered by trees. They then estimate the surface area of a leaf and how many leaves are on the typical tree of a particular type. After determining how many trees there are on Murre Island, students figure out for how many people the trees can provide oxygen.

Students compare data on ancient forest protection for 12 national forests. They compare the numbers of square kilometers and the percents of ancient forests that are protected. They then compare the protection of ancient forests on Murre Island with the protection of these 12 national forests.

Goals

Students will:

- organize and describe data;*
- use statistical summaries and graphs;*
- calculate with fractions, decimals, and percents;
- read, interpret, and use information from maps or graphical displays of data;
- choose and use appropriate techniques to analyze data;*
- make inferences based on data; *
- make decisions based on statistical information and justify these decisions.*

** These goals are addressed in this section and are assessed in other sections of the unit.*

Pacing

- approximately three 45-minute class sessions

About the Mathematics

In this section, students use a grid for estimating the area of an irregular shape. The shape is placed on a grid and the area of the shape is found by counting the number of squares covered by it. When finding area this way, students should round to the nearest $\frac{1}{2}$ square for estimating purposes. If the squares on the grid are 1 centimeter by 1 centimeter, the area is expressed in square centimeters. In order to convert square centimeters to square meters, students should think of a square that is covered by 100 rows of 100 square centimeters each. The square is 1 meter by 1 meter; therefore, 1 square meter is equal to 10,000 square centimeters.

Students also work with part-whole relationships and make absolute and relative comparisons. Data are presented in a bar graph. Students use the bar graph to estimate the percent in a particular category. They also use the bar graph to make absolute and relative comparisons.

Materials

- Student Activity Sheets 1, 6, and 9, pages 121, 126, and 129 of the Teacher Guide (one of each per student)
- Student Activity Sheets 7–8, pages 127–128 of the Teacher Guide, optional (one of each per student)
- calculators, pages 37, 39, 41, 43, and 45 of the Teacher Guide, (one per student)
- leaves, page 39 of the Teacher Guide, optional (five leaves per student)

Planning Instruction

You may start this section by asking your students whether they have ever been to a national forest. Ask: *Why is it important to protect trees in a national forest? How are trees beneficial? Are some kinds of trees more beneficial than others?*

Students can work on problems 1 and 2 and 7–11 individually or in small groups. Problems 3–6 and 12 may be completed in small groups. Problems 13 and 14 may be completed individually.

There are no optional problems in this section.

Homework

Problems 13 and 14 (page 46 of the Teacher Guide) can be assigned as homework. Students' responses to these problems may also be assigned as Writing Opportunities.

Planning Assessment

- Problem 2 can be used to informally assess students' ability to calculate with fractions, decimals, and percents.
- Problem 11 can be used to informally assess students' ability to read, interpret, and use information from maps or graphical displays of data.

B. DATA ABOUT ENDANGERED FORESTS

Murre Island Forests

Murre Island is home to a variety of species of wildlife, including the northern spotted owl. Since most of the forests in the Pacific Northwest in which this owl lives have been cut down for timber, the bird is now an endangered species.

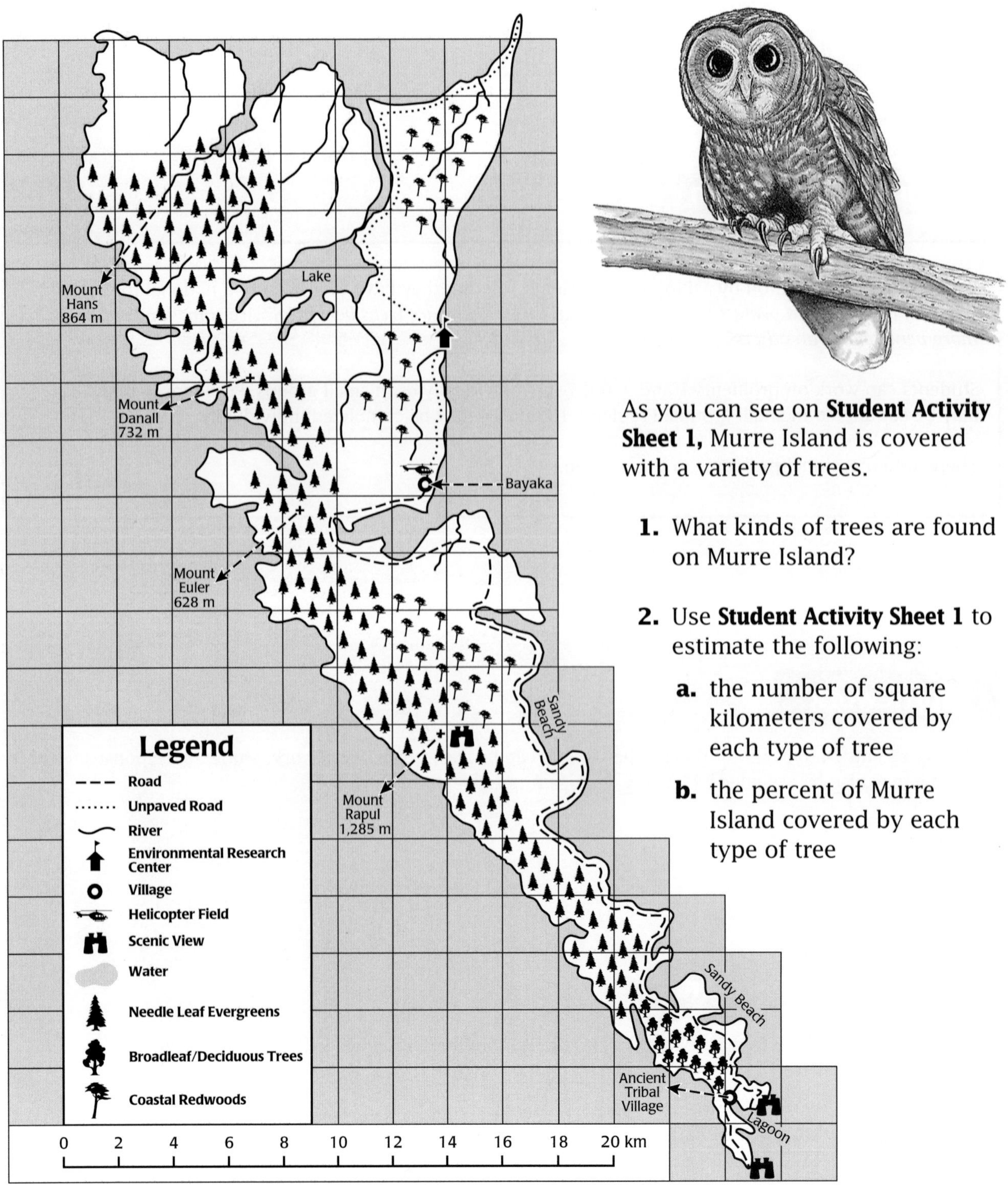

As you can see on **Student Activity Sheet 1,** Murre Island is covered with a variety of trees.

1. What kinds of trees are found on Murre Island?

2. Use **Student Activity Sheet 1** to estimate the following:

   a. the number of square kilometers covered by each type of tree

   b. the percent of Murre Island covered by each type of tree

Solutions and Samples

of student work

1. Trees on Murre Island:
   - needle leaf evergreens
   - broadleaf/deciduous trees
   - coastal redwoods

2. Answers may vary. Answers should be close to the answers provided below. *Note:* Each square on the grid represents an area of 4 square kilometers.

   a. • Needle leaf evergreens cover about $22\frac{1}{2}$ grid squares; $22\frac{1}{2} \times 4 \text{ km}^2 \approx 90 \text{ km}^2$.
   - Broadleaf/deciduous trees cover about $1\frac{1}{2}$ grid squares; $1\frac{1}{2} \times 4 \text{ km}^2 = 6 \text{ km}^2$.
   - Coastal redwoods cover about $7\frac{1}{2}$ grid squares; $7\frac{1}{2} \times 4 \text{ km}^2 \approx 30 \text{ km}^2$.

   b. Murre Island covers about 68 grid squares; $68 \times 4 \text{ km}^2 = 272 \text{ km}^2$.
   - Needle leaf evergreens cover about 33% of Murre Island; 90 km^2 out of 272 km^2 is about 90 out of 270, which is $\frac{1}{3}$ or 33%.
   - Broadleaf/deciduous trees cover about 2% of Murre Island; 6 km^2 out of 272 km^2 is about 6 out of 270, which is about 1 out of 45, or just a little more than 2%.
   - Coastal redwoods cover about 11% of Murre Island; 30 km^2 out of 272 km^2 is about 30 out of 270, which is $\frac{1}{9}$ or about 11%.

Hints and Comments

Materials Student Activity Sheet 1 (one per student); calculators (one per student)

Overview Students study a land-area map of Murre Island in order to estimate the number of square kilometers and the percent of Murre Island covered by each type of tree.

About the Mathematics On this page, students estimate the areas of irregular shapes using the method that was introduced in the unit *Reallotment.* They find the area by estimating the number of squares covered by the shape.

Students also use the relationships between ratios, fractions, decimals, and percents to calculate percents.

Planning Students may work in small groups or individually on problems **1** and **2.** Discuss students' strategies for estimating the areas and the percents for problem **2**. Problem **2** may be used as an informal assessment.

Comments about the Problems

1. You may want to discuss the tree types listed in the map legend. The following trees might be found on Murre Island:
   - Needle leaf evergreens: Douglas firs, spruces, and pines
   - Broadleaf/deciduous trees: Oregon (bigleaf) maples and red alders
   - Coastal redwoods: giant sequoias and redwoods, which are also evergreens

2. **Informal Assessment** This problem assesses students' ability to calculate with fractions, decimals, and percents.

   a. Have students estimate the area of Murre Island covered by trees using the method of counting grid squares. You may want to have students round their estimates to the nearest $\frac{1}{2}$ square for easier calculations.

   b. Allow students to use their own strategies for finding the percents.

Activity

One reason that trees are important is that they release oxygen. The amount of oxygen that a tree releases depends on the total area of its leaves.

- Find a tree near your school or in your neighborhood that has broad leaves.
- Estimate the number of leaves on the tree.
- Collect five average-sized leaves from the tree (be careful not to damage the tree).

Oregon maple and red alder are two broadleaf, deciduous trees found on Murre Island.

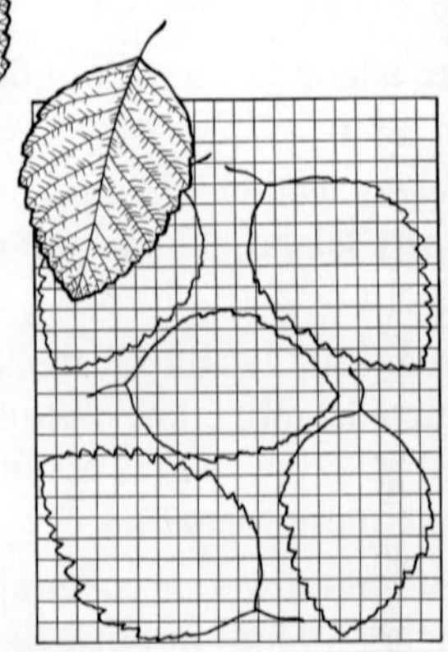

3. a. Trace the leaves onto the grid on **Student Activity Sheet 6** and then estimate the average surface area of a single leaf.

b. What is the total leaf area of your tree?

Suppose biologists calculate that a leaf area of 25 square meters produces enough oxygen on a sunny day for one person.

4. For how many people does your tree supply oxygen on a sunny day?

5. If there is about one tree per 20 square meters in the forested areas of Murre Island, estimate the total number of broadleaved trees on the island. Explain how you found your answer.

6. Assume that the total leaf area of each broadleaved tree on Murre Island is the same as the total leaf area of your tree. For how many people do all of the broadleaved trees on Murre Island supply oxygen on a sunny day?

Solutions and Samples
of student work

3. **a.** Estimates will vary, depending on the size and type of tree. Sample estimates:

   Using Student Activity Sheet 7, the average size of a red alder leaf is 58 cm^2. The approximate areas for the leaves are 52 cm^2 (center), 53 cm^2 (top left), 55 cm^2 (bottom right), 60 cm^2 (top right), and 71 cm^2 (bottom left).

   Using Student Activity Sheet 8, the size of the Oregon maple leaf is 206 cm^2.

   b. Estimates will vary, depending on the size and type of tree. Sample estimates:

   The total leaf area for a red alder tree is about 11.6 m^2. A red alder tree has roughly 2,000 leaves. So, the total leaf area is equal to 58 $cm^2 \times$ 2,000 leaves = 116,000 cm^2 = 11.6 m^2.

   The total leaf area for an Oregon maple is about 24.7 m^2. An Oregon maple has roughly 1,200 leaves. So, the total leaf area is equal to 206 $cm^2 \times$ 1,200 leaves = 247,200 cm^2 = 24.7 m^2.

4. Answers will vary, depending on the size and type of tree. If the total leaf area is expressed in square meters, students may divide the leaf area by 25. Using the example of a red alder tree from Student Activity Sheet 7, 11.6 $m^2 \div 25 = 0.5$; about 0.5 people are supplied with oxygen per day. In other words, the tree could supply about half of the oxygen needed for one person in one day.

5. 300,000 broadleaved trees. Explanations will vary. Sample explanation:

   About 6 km^2 of Murre Island are covered with broadleaved trees. Since there is about 1 tree per 20 m^2, divide the total number of square meters by 20.

   6 km^2 = 6,000,000 m^2; 6,000,000 ÷ 20 = 300,000 trees

6. Use the number of people supplied with oxygen from problem **4** and multiply by the number of broadleaved trees from problem **5.**

   Using the example of a red alder tree and 1 tree per 20 m^2, 0.5 people × 300,000 trees = 150,000 people supplied with oxygen.

Hints and Comments

Materials Student Activity Sheet 6 (one per student); Student Activity Sheets 7 and 8, optional (one per student or group); calculators (one per student); leaves, optional (five leaves per student)

Overview Students collect five leaves from one tree and estimate the surface area of an average leaf. They then estimate how many leaves are on the tree. Students also determine for how many people one tree supplies oxygen and for how many people all of the trees on Murre Island supply oxygen.

About the Mathematics In problem **3,** students estimate the area of an irregular shape using a grid the same way that they did in problem **2** on Student Book page 15. For problems **4** and **6,** students use the method for converting from square centimeters to square meters that was developed in the unit *Reallotment.* They visualize 1 square meter as a 1-by-1-meter square. Since 1 meter equals 100 centimeters, 1 square meter is also equivalent to a 100-by-100-centimeter square. Therefore, 1 square meter is the same as 10,000 (100 × 100) square centimeters.

Planning Any type of broadleaved tree can be selected for the activity, as long as its leaves will fit on Student Activity Sheet 6. Before completing the activity, ask students to develop a strategy to estimate the number of leaves on a tree. Rather than have students collect real leaves, you might have students use Student Activity Sheets 7 and 8, which show tracings of leaves. Students may work in small groups on problems **3–6.**

Comments about the Problems

3. Have students share their results and note the differences in area based on the different kinds of broadleaved trees that they have found.

5. You may want to have students estimate the average distance between trees in a forested area. Students could go to a nearby wooded area to find their estimates, or they could visualize their classroom filled with trees and determine how many trees would fit and the area of the classroom. You may use the number of trees per square meter determined by your students instead of the given measure, one tree per 20 square meters. (In a dense forest it is not unusual to have one tree per 10 square meters.)

The broadleaf, deciduous forest on Murre Island was once all coastal redwoods. Most of the old coastal redwoods were cut down for timber during the last 30 years.

7. What are some advantages of cutting down coastal redwoods? some disadvantages?

The Environmental Research Center has calculated that, compared to 1965, only 24 percent of the old coastal redwoods remain on Murre Island.

8. a. Estimate the area that was covered by coastal redwoods in 1965.

b. On **Student Activity Sheet 9,** draw the coastal redwood forests as they might have appeared in 1965. (Remember that the broadleaf, deciduous forest was then a coastal redwood forest.)

ESS wants to protect the remaining coastal redwoods on Murre Island, as well as 80 percent of the other ancient needle leaf forests.

9. How will this forest protection affect the development of Murre Island?

Solutions and Samples
of student work

7. Answers will vary. Sample response:

   Some advantages of cutting down coastal redwoods are acquiring timber and space for other uses.

   Some disadvantages are loss of habitat, erosion of the soil, less protection from winds, fewer birds and animals, loss of the beauty of the trees, damage to the watershed, and loss of a source of oxygen.

8. **a.** Coastal redwoods covered about 120 km^2 in 1965. Strategies will vary. Sample strategies:

   Coastal redwoods currently cover about 30 km^2, which was calculated in problem **2a.** This is 24% of the area that coastal redwoods covered in or before 1965; 24% is about $\frac{1}{4}$, and 30 km^2 is $\frac{1}{4}$ of 120 km^2.

   Using a bar:

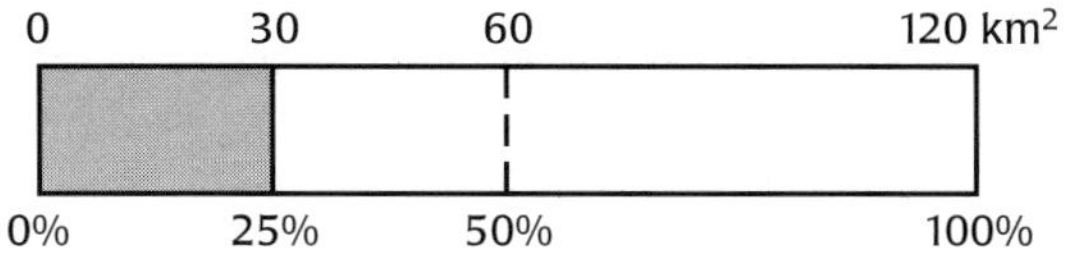

   b. Drawings will vary. Students should draw an additional 90 km^2 of coastal redwoods, to make a total of 120 km^2 (30 squares).

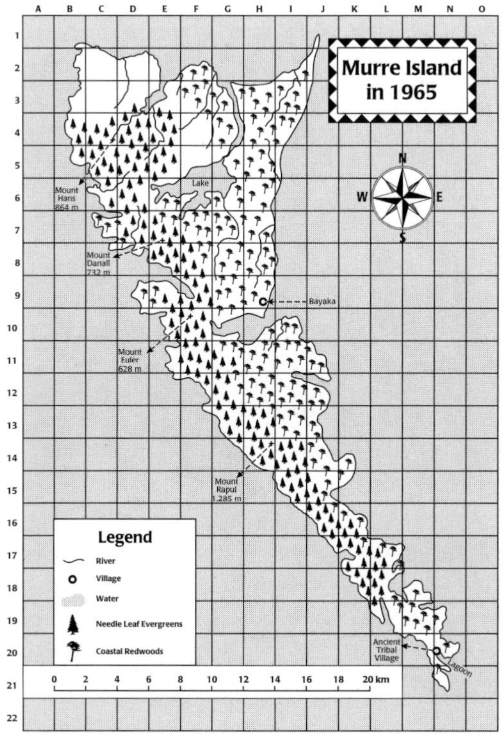

9. Answers will vary. Sample response:

   The development of Murre Island will be limited because very few trees can be cut down.

Hints and Comments

Materials Student Activity Sheet 9 (one per student); calculators (one per student)

Overview Students investigate how the area of a forest may change over a period of time. They estimate the area of Murre Island that was covered by coastal redwoods in 1965 and draw a new map showing this area.

About the Mathematics In problem **8,** students work with part-whole relationships. The part (24% of the area) is given, and students are asked to find the whole (100% of the area). Visual models such as a percent bar, which was introduced in the unit *Per Sense* and further developed in *More or Less,* could be used if students are still having trouble understanding the part-whole relationship.

Planning Students may work individually or in small groups on problems **7–9.** Discuss students' strategies for problem **8.**

Comments about the Problems

8. Allow students to use their own strategies. Students should be able to explain how they arrived at their answers.

National Forests

Plans to protect ancient forests on Murre Island are typical of efforts in other parts of the United States. The table on the right shows information about 12 national forests in the Pacific Northwest.

National Forest Data

Name	Size of Ancient Forest (in km^2)	Area Protected (in km^2)	Area Unprotected (in km^2)
Mt. Baker-Snoqualmie	2,370	1,698	672
Olympic	767	603	164
Gifford Pinchot	1,384	771	613
Mt. Hood	1,354	579	775
Willamete	2,449	1,361	1,088
Siuslaw	541	299	242
Umpqua	1,240	386	854
Rogue River	607	306	301
Siskiyou	1,346	486	860
Klamath	1,293	710	583
Six Rivers	832	365	467
Shasta-Trinity	1,105	629	476

10. Organize the data in the table and then compare the protection of ancient forests in these 12 national forests.

Maria uses the information in the above table to make a graph. She presents her graph to the members of ESS.

11. Estimate the percent of the ancient forest in the Mt. Baker-Snoqualmie National Forest that is unprotected.

10. The data may be organized in two ways. If the data are ranked according to square kilometers protected, then the order is the following:

Name	Size of Ancient Forest (in km²)	Area Protected (in km²)	Area Unprotected (in km²)
Mt. Baker-Snoqualmie	2,370	1,698	672
Willamete	2,449	1,361	1,088
Gifford Pinchot	1,384	771	613
Klamath	1,293	710	583
Shasta-Trinity	1,105	629	476
Olympic	767	603	164
Mt. Hood	1,354	579	775
Siskiyou	1,346	486	860
Umpqua	1,240	386	854
Six Rivers	832	365	467
Rogue River	607	306	301
Siuslaw	541	299	242

If the data are ranked according to the percent of ancient forest protected, then the order is the following:

Name	Size of Ancient Forest (in km²)	Area Protected (in km²)	Percent Protected
Olympic	767	603	78.6
Mt. Baker-Snoqualmie	2,370	1,698	71.6
Shasta-Trinity	1,105	629	56.9
Gifford Pinchot	1,384	771	55.7
Willamete	2,449	1,361	55.6
Siuslaw	541	299	55.3
Klamath	1,293	710	54.9
Rogue River	607	306	50.4
Six Rivers	832	365	43.9
Mt. Hood	1,354	579	42.8
Siskiyou	1,346	486	36.1
Umpqua	1,240	386	31.1

11. about 28% (accept answers between 25% and 35%)

Materials calculators (one per student)

Overview Students reorganize and compare data from a table. They then estimate the percent of an ancient forest that is unprotected using the data from a bar graph.

About the Mathematics For problem **10,** students may make either absolute or relative comparisons. They may compare only the sizes of the protected areas (an absolute comparison) or they may compare the relationship of the protected area to the total area of each forest (a relative comparison). Absolute and relative comparisons are introduced in the unit *Ratios and Rates* and further developed in *Cereal Numbers.*

The data on this page are also presented in a bar graph. The bars are segmented, or stacked, and they display the distribution of the categories (area protected and area unprotected) for each item (each national forest). Students may use the bar graph to make both absolute and relative comparisons.

Planning Students may work individually or in small groups on problems **10** and **11.** Discuss students' solutions and strategies for problems **10** and **11** after they finish problem **12** on the next page (Student Book page 19). Problem **11** may be used as an informal assessment.

Comments about the Problems

10. Discuss the advantages of each ordering scheme that students develop along with the two displays on Student Book page 18. For example, the advantage of the table on Student Book page 18 is that it shows the exact number of square kilometers that are protected in each national forest. In the bar graph, the emphasis is on the portion of each forest that is protected.

11. **Informal Assessment** This problem assesses students' ability to read, interpret, and use information from maps or graphical displays of data.

One member of ESS makes the following comment about Maria's graph:

"Klamath National Forest does a better job protecting its ancient forests than Siuslaw. In Klamath, 710 square kilometers of ancient forest are protected; in Siuslaw, only 299 square kilometers of ancient forest are protected."

Another ESS member makes the following claim:

"Klamath and Siuslaw both provide good protection of their ancient forests because 55 percent of the ancient forests in both national forests are in protected areas."

12. **a.** Who do you think is right? Explain.

b. Which method of comparing do you prefer? Why?

c. Which national forest do you think is doing the best job of protecting ancient forests? Why?

Solutions and Samples
of student work

12. a. Both claims are correct. Explanations will vary. Sample explanations:

- The first statement is correct because 710 km^2 is more than 299 km^2.
- The second statement is correct because Klamath and Siuslaw protect the same percent of the total area of ancient forest.

b. Answers will vary. Sample response:

I prefer comparing the percents. The amount of ancient forest that is protected should depend on the size of the total area of ancient forest.

c. Answers and explanations will vary. Sample response:

Mt. Baker-Snoqualmie is doing the best job protecting its ancient forests because it has the largest area of protected ancient forest (1,698 km^2) and the second highest percent of protected ancient forest (71.6%).

Hints and Comments

Materials calculators (one per student)

Overview Students reflect on different ways of comparing data.

Planning Students may work in small groups on problem **12.** You might want to look back to problem **10** and discuss the different ways of making comparisons. Also discuss students' solutions and strategies for problem **12.**

Comments about the Problems

12. c. If students made relative comparisons in problem **10,** they can use these results to answer problem **12c.**

If students are having difficulty making relative comparisons, encourage them to study the bar graph on the previous page (Student Book page 18) and estimate the shaded portion of the bar.

Summary

In order to help ESS make decisions about the forested areas of Murre Island, you made several estimates that are difficult or impossible to actually measure. For example, you estimated the area of Murre Island that is covered by each type of tree. You also estimated the total leaf area of a tree, the total number of broadleaf trees on Murre Island, and the number of people for whom these trees would provide oxygen.

In addition, you compared the protected areas of several national forests by examining the number of square kilometers of ancient forest that are protected and the percent of ancient forest that is protected.

Summary Questions

13. ESS has decided to protect all coastal redwoods and 80 percent of the ancient needle leaf trees on Murre Island. Draw a bar for Murre Island like those shown on Maria's graph on page 18.

14. Compare the protection of ancient forests on Murre Island with the protection of national forests on Maria's graph. What do you think of ESS's protection plan for Murre Island? Explain.

13. **Protection of Ancient Forests**

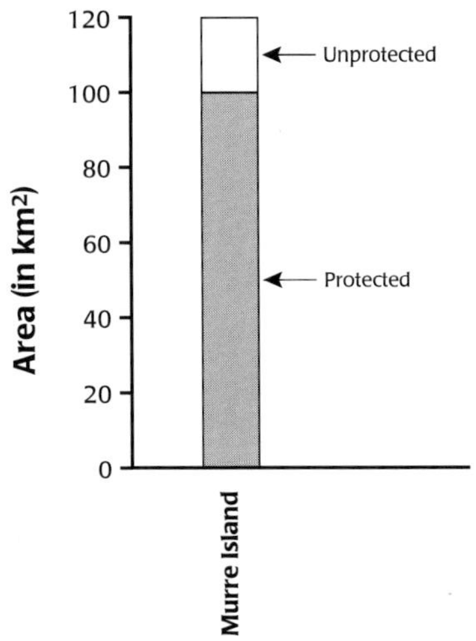

The needle leaf evergreens and the coastal redwoods make up the ancient forests on Murre Island. The needle leaf evergreens cover 90 km^2 and the coastal redwoods cover 30 km^2, so the ancient forests cover a total of 120 km^2.

The total area protected consists of 100% of all coastal redwoods, or 30 km^2, and 80% of needle leaf forest: 80% of 90 km^2 = 72 km^2. So 102 km^2 of ancient forest out of 120 km^2 will be protected on Murre Island.

14. Answers will vary. Sample response:

Since about 85% of the ancient forests on Murre Island will be protected, Murre Island would rank first in a relative comparison with the national forests on Maria's graph. ESS has an excellent protection plan for Murre Island.

Overview Students draw a bar similar to those on the bar graph on Student Book page 18 to show the percent of ancient forest on Murre Island that is protected. They then compare the percent of protected forest on Murre Island with the percents of other national forests.

Planning Students may work individually on problems **13** and **14.** These problems may be assigned as homework.

Comments about the Problems

13–14. These problems may be assigned as homework.

13. Students should use their answers from problem **2** on Student Book page 15 to find the areas covered by needle leaf evergreens (90 square kilometers) and coastal redwoods (30 square kilometers), as well as the entire forested area (126 square kilometers).

Some students may draw the bar for Murre Island on the same scale as the bars on Maria's bar graph, shown on Student Book page 18. In this case, the bar will be very small. You may suggest that students change the scale; the relative comparison does not depend upon the scale of the bar.

Writing Opportunity Students may write their answers to problems **13** and **14** in their journals.

Work Students Do

Students analyze the data collected in their surveys from Section A. They organize, summarize, and graphically represent the class data. They discuss whether their sample was random and representative of a given population. They then use the data to write statements that may be useful in planning tourist activities on Murre Island, taking into consideration the protection of the environment. Then students study the results of a survey and use them to make decisions about the development of Murre Island.

Goals

Students will:

- organize and describe data;
- use statistical summaries and graphs;
- calculate with fractions, decimals, and percents;
- judge whether or not samples are representative and random;
- read, interpret, and use information from maps or graphical displays of data;*
- choose and use appropriate techniques to analyze data;
- make inferences based on data;
- make decisions based on statistical information and justify these decisions.*

** These goals are addressed in this section and are assessed in other sections of the unit.*

Pacing

- approximately four 45-minute class sessions

Vocabulary

- random sample
- statistical summary

About the Mathematics

Statistics and graphs work together in any data analysis. The graphs provide a visual picture of the data, and the statistics summarize a characteristic of the data: they can tell you about typical values or about how spread out the data are. Graphical representations of data can delineate a center, the overall shape of the data, gaps, clusters, and outliers. A statistic is a number that summarizes a set of data. It may be a count, a representative value in a set of data, or an arithmetic combination of data points. Mean, median, and mode are examples of these. The mode is the most frequent data point that appears in the data set. The median and the mean are used to describe the "center." The median is the data point in the middle when the data are ordered. It is not influenced by outliers in the data set. The mean is computed from the actual values of the data (their sum divided by the number of values) and is influenced by outliers in the data set. It is important to consider the range and the spread of the data when giving the mean. A good description of data should include the range and the mean.

Materials

- completed questionnaires from Section A, pages 51, 53, and 55 of the Teacher Guide
- newspapers and magazines, page 61 of the Teacher Guide, optional (several of each per class)
- scissors, page 61 of the Teacher Guide, optional (one pair per student)
- drawing paper, page 61 of the Teacher Guide, optional (one sheet per student)
- glue or tape, page 61 of the Teacher Guide, optional (one bottle or roll per group)

Planning Instruction

Begin this section by giving groups time to tabulate the ages of the people they surveyed. Ask: *Why might it be important to know the age distribution of a survey sample?* [to determine whether the sample is representative of a population and to explain the trends in the results, among other reasons]

Displaying the results from individual groups on a particular survey item alongside the results from the entire class on the same item will allow students to see how the results change as the sample gets larger. If you teach multiple classes, the results of a particular survey item for all students can also be displayed.

Students may work on problems 8, 11, 16, and 17 individually. You may want students to work individually or in small groups on problems 12–15. Problems 3–5, 7, and 9 may be completed in small groups or as whole-class activities. The remaining problems may be completed in small groups.

Problems from this section, which includes the analysis of survey data, can be completed after Section D and can be used as a final assessment of the unit.

Homework

Problems 8, 10, and 11 (page 54 of the Teacher Guide) and problems 16 and 17 (page 60 of the Teacher Guide) can be assigned as homework. The Extensions (pages 53 and 61 of the Teacher Guide) and the Writing Opportunity (page 55 of the Teacher Guide) may also be assigned as homework.

Planning Assessment

- Problem 1 can be used to informally assess students' ability to organize and describe data and to use statistical summaries and graphs.
- Problems 4b, 5, 14, and 15 can be used to informally assess students' ability to judge whether or not samples are representative and random.
- Problem 10 can be used to informally assess students' ability to calculate percents.
- Problem 11 can be used to informally assess students' ability to organize and describe data, to choose and use appropriate techniques to analyze data, and to make inferences based on data.
- Problem 13 can be used to informally assess students' ability to use statistical summaries and graphs and to make inferences based on data.
- Problems 16 and 17 can be used to informally assess students' ability to calculate with fractions, decimals, and percents.

C. INTERPRETING DATA

STATISTICAL SUMMARIES

In Section A, you conducted a survey about how people prefer to spend their vacations. You are now going to tabulate the results and write statistical summaries about the information you gathered. In a *statistical summary,* you use graphs (such as histograms, bar charts, pie charts, or box plots) and numbers (such as the range, mean, median, or mode) to summarize and interpret survey results.

Put all of the completed surveys from your group together.

1. Write a statistical summary of the ages of the people who were surveyed by your group. Be prepared to present your results to the class.

Solutions and Samples
of student work

1. Summaries will vary. Sample summaries are given below, based on a sample of 30 people with the following ages:

10	12	42
11	21	43
11	30	44
11	35	44
11	36	44
11	37	47
11	38	49
11	40	49
12	41	52
12	41	57

- The range is 10–57 years of age. The mean age is 30.4. The median age is 36.5. The mode is 11.
- Using a stem-and-leaf plot:

Stem	Leaves
0	
1	0 1 1 1 1 1 1 1 2 2 2
2	1
3	0 5 6 7 8
4	0 1 1 2 3 4 4 4 7 9 9
5	2 7
6	
7	

- Using a box plot:

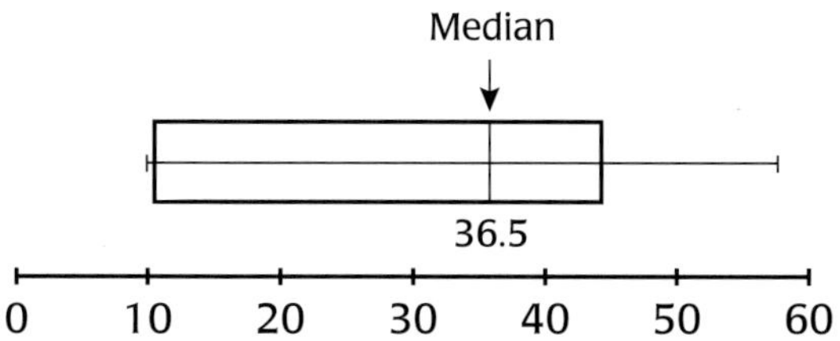

The stem-and-leaf plot shows clusters in the teens, thirties, and forties. The box plot indicates that 50% of the people were between 11 and 44 years old and 25% were between 36.5 and 44 years old.

Hints and Comments

Materials completed questionnaires from Section A

Overview In this section, students begin to analyze the data they collected in their survey from Section A.

Planning Divide the students into small groups of about four so that each group has 16–40 completed questionnaires to pool. Have students label each form with the name or number of the group to which it belongs. Each group should keep their surveys for reference throughout the rest of the unit.

Encourage students to use the one-number summaries and graphs developed in *Dealing with Data:* mean, median, mode, and range; number-line plots, stem-and-leaf plots, histograms, and box plots. Pie charts were developed in *Fraction Times.*

Problem **1** can be used as an informal assessment.

Comments about the Problems

1. Informal Assessment This problem assesses students' ability to organize and describe data and to use statistical summaries and graphs.

If students use the same categories as the graph on Student Book page 22, the questions that follow will be easier to answer. If you notice that there is not much variety in students' graphical representations, you may suggest other ideas (bar graphs, box plots, tables, and so on). Discuss the advantages and disadvantages of different graphical representations. Discuss which statistical measurements together give the best picture of the data. (See the introductory pages of this section for information about these points.)

The class might discuss using common scales for the plots and the same radius for pie charts. This will help in comparing the results from different groups.

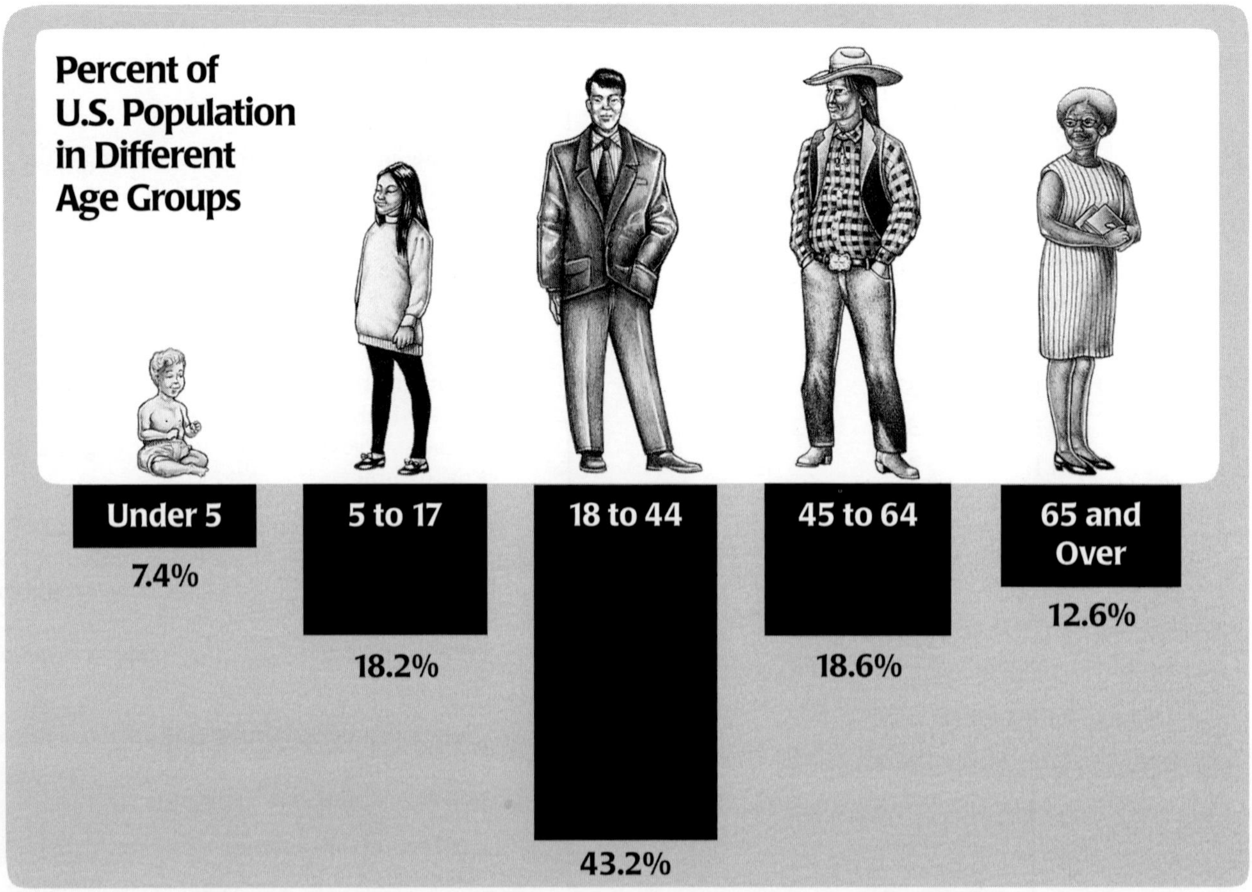

Source: Data from *The World Almanac and Book of Facts 1996* (Mahwah, New Jersey: Funk and Wagnalls Corporation, 1995), 378.

The above chart shows U.S. population distribution by age in 1990 (based on the 1990 census).

2. If there were 200 people in a sample, how many of them would you expect to be older than 65?

3. Write a statistical summary of the ages of all people surveyed by your class. (*Note:* You want to be able to compare the age distribution of the people who were surveyed with the age distribution of the population of the United States in 1990.)

4. a. Are the ages of the people in the class sample distributed in the same way that the ages of the overall population are distributed?

b. Is the class sample representative of the overall population? Explain.

A *random sample* is chosen from a population in such a way that everyone in the population has an equal chance of being selected for the sample.

5. Is the sample that your class took for the survey a random sample? Why or why not?

2. You would expect about 25 (12.6% of 200) people to be 65 and over.

3. Summaries will vary, depending on the class survey results. Sample strategy based on the sample survey shown in the solutions column for problem **1:**

   First determine the number of people in each category and then calculate the percent of people in each category.

Age Group	Number	Percent
Under 5	0	0
5 to 17	11	37
18 to 44	14	47
45 to 64	5	17
65 and over	0	0

   (*Note:* The percents are rounded to the nearest whole number.)

4. **a.** Answers will vary, depending on the class sample. The distributions are probably different.

   b. Answers will vary. The class sample is probably not representative of the overall population because the sample was not random and because people from only a single area were sampled, among other reasons.

5. Students' samples are most likely not random. They are probably biased towards their friends, family members, classmates, and so on.

Materials completed questionnaires from Section A

Overview Students write a statistical summary of the ages of all the people who were surveyed. They discuss whether their sample is random and representative.

About the Mathematics Before sampling, you must define your population. The definition of a population is determined by the information you want to gather. For example, if you want to know which books middle school students like to read, the results of a survey of only seventh-graders will inaccurately represent the total population.

Planning Discuss students' answers and solutions to problems **2–4a** before they continue working on problems **4b** and **5.** Problems **4b** and **5** can be used as informal assessments. Students may work on problems **2–5** in small groups. Problems **3–5** may also be completed as a whole-class activity.

Comments about the Problems

2. The mathematical answer of 25.2 doesn't make sense when referring to people.

3. Students should use the same age categories as in the graph on Student Book page 22. The effect of sample size can be discussed as students compare the results from their small groups with the class results.

4b–5. Informal Assessment These problems can be used to assess students' ability to judge whether or not samples are representative and random.

5. The answer to this question depends on how students define their populations.

Extension You may want to discuss whether it is possible to use the results of problem **1** (small groups) to draw conclusions about the class data as a whole. You may not be able to find the mean age of the whole group by using the means of the smaller groups because all the groups may not have the same number of surveys.

Vacation Activities

You are now going to take a closer look at the activities people participate in while on vacation. ESS thinks that there are major differences between what adults and children prefer to do on vacation. ESS suggests marking each survey with the following code:

A = adults

C = children (under age 18)

TG = travels with tour group

FM = travels with family

FR = travels with friends

Mark each survey response from your group with an A or a C and a TG, an FM, or an FR.

6. How many people did your group survey in each category?

7. Make a table like the one below to show the results of the entire class.

	Adults	Children	Total
Travels with Tour Group			
Travels with Family			
Travels with Friends			
Total			

8. Do adults and children travel in different types of groups? Explain.

9. Make a table like the one on the right using the information collected about activities from the surveys of everyone in the class.

Activity	Adults	Children

10. Do children and adults prefer different activities while on vacation? Be prepared to explain how you reached your conclusions.

11. Based on the results of the class surveys, how should ESS develop Murre Island so that it is attractive to visitors?

Solutions and Samples
of student work

6. Answers will vary, depending on the class survey data.

7. Tables will vary. Sample table:

	Adults	Children	Total
Travels with Tour Group	20	2	22
Travels with Family	97	80	177
Travels with Friends	23	3	26
Total	140	85	225

8. Answers will vary. Most likely, children travel more with families. In the sample data given for problem **7,** above, 94% of the children travel with family while only 69% of the adults do.

9. Tables will vary, depending on the class survey data. Sample table:

Activity	Adults	Children
Swimming	107	82
Hiking	93	60
Sailing	89	55
Bicycling	88	78

10. Answers will vary. Students can use percents in their explanations or draw bar graphs, pie charts, or even a scatter plot with Percents of Adults on one axis and Percents of Children on the other and with each activity represented by a dot and labeled with an abbreviation. A diagonal can be drawn on the scatter plot to show equal percents of children and adults.

11. Answers will vary, depending on students' data. Recommendations may include the following:
    - building a lodge/hotel with guest rooms that are geared toward families,
    - building as many hiking and mountain biking trails as possible,
    - providing for swimming and sailboat rentals at the sandy beaches on the island,
    - hiring guides for tour groups.

Hints and Comments

Materials completed questionnaires from Section A

Overview Students analyze the results of their survey.

Planning Students may work on problems **6–10** in the same small groups. Students may work on problem **11** individually. In problems **7** and **9,** students have to compile the data of the entire class. The easiest way to organize this is to have the groups first tabulate their own results and then collect the information from other groups, or you may prefer to fill in the table as a whole-class activity. Problems **8, 10,** and **11** can be assigned as homework. Problems **10** and **11** can also be used as informal assessment.

Comments about the Problems

6. Students should first divide the forms into two subgroups (children and adults) and then sort each subgroup according to the type of traveling companions preferred.

8. Homework This problem may be assigned as homework. Students will need to use fractions or percents to make comparisons.

9. It is important to take into account that those surveyed were allowed to choose more than one activity.

10–11. Homework These problems may be assigned as homework.

10. Informal Assessment This problem assesses students' ability to calculate percents. Students will need to use fractions or percents to make comparisons.

11. Informal Assessment This problem assesses students' ability to organize and describe data, to choose and use appropriate techniques to analyze data, and to make inferences based on data. It is important that students support their recommendations with data and graphical representations.

Writing Opportunity You may have students write their responses to problem **11** in their journals.

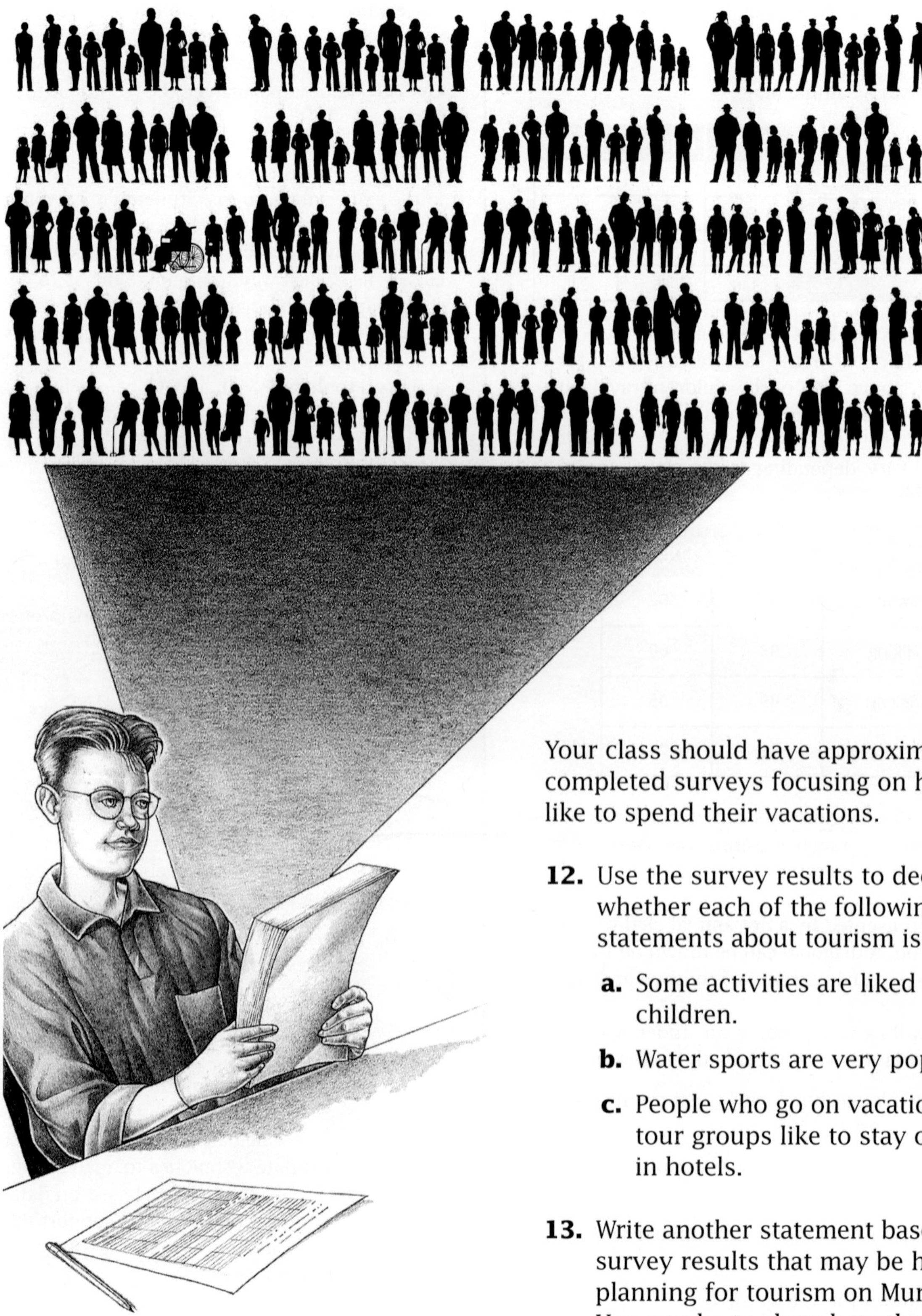

Your class should have approximately 200 completed surveys focusing on how people like to spend their vacations.

12. Use the survey results to decide whether each of the following statements about tourism is accurate:

a. Some activities are liked only by children.

b. Water sports are very popular.

c. People who go on vacation with tour groups like to stay overnight in hotels.

13. Write another statement based on your survey results that may be helpful for planning for tourism on Murre Island. Use graphs to show how the data from the surveys support your statement.

Solutions and Samples
of student work

12. Answers will vary, depending on students' data.

a–b. Answers may vary. The representation and explanation from problems **9** and **10** can be used to make statements and to justify them.

c. Answers will vary. Answers to two different questions on the questionnaire must be combined in order to draw conclusions.

13. Statements and graphs will vary. Sample statements:

Adults who travel with friends are the most active people. They bicycle, swim, canoe, climb mountains, and go horseback riding.

Twenty percent of the families surveyed like to camp. All of these families fish.

Hints and Comments

Overview Students further organize and analyze the results of their survey. They use their results to decide whether some given statements are accurate.

Planning Students may work on problems **12** and **13** individually or in small groups. Problem **13** can be used as an informal assessment.

Comments about the Problems

12. It is not necessary that each student or even each small group investigate all of the statements in problem **12.** Students can choose among them, or you can assign them to different groups.

Students must have access to the class data that apply to the statement in question. For statements **a** and **c,** the table from problem **9** will suffice. For statement **b,** groups will have to provide each other with a breakdown of where people who travel with tour groups like to stay overnight.

13. Informal Assessment This problem assesses students' ability to use statistical summaries and graphs and to make inferences based on data.

Students must be careful making statements that describe tourist preferences based on their class survey. The survey results may reflect the community from which the sample was drawn but may not be reflective of other communities. Students should support their answers with statistics or graphs.

14. Is it appropriate to use your data set to make decisions for the development of Murre Island? Why or why not?

15. ESS has distributed questionnaires to a random sample of people in the United States.

a. What factors must be considered when choosing a random sample?

b. Why is it important to have a random sample?

c. How do you think ESS selected a random sample of people in the United States?

Solutions and Samples
of student work

14. No, the class data is not appropriate for making decisions about Murre Island. The class data probably is not a representative sample of the population of the United States.

15. a. Answers may vary. The sample should represent the different kinds of people who live in the United States (people of different ages, social classes, regions, ethnic origins, genders, sexual orientations, and so on).

b. Answers may vary. Students should note that samples obtained in a way that is not random may be biased and can generate misleading conclusions about the population. If a sample includes only people from Florida, for instance, data on favorite vacation activities may be very different from the data that would be given by people in Utah.

c. Answers will vary. Students may suggest various plans, such as using random phone digits to call people or traveling to random grocery stores across the country to survey customers.

Hints and Comments

Overview Students decide whether their survey sample is representative of prospective Murre Island tourists. Then they discuss conditions for taking a random sample.

About the Mathematics A random sample is one in which every member of the defined population has the same chance of being included. One way to do this is by assigning each member of the population a number, physically mixing the numbers, and then drawing a sample. The mixing must be done in such a way that every number has an equal probability of being drawn.

Planning Students may work on problems **14** and **15** individually or in small groups. These problems can be used as informal assessments.

Comments about the Problems

14–15. Informal Assessment These problems assess students' ability to judge whether or not samples are representative and random.

Summary

In this section, you used statistical summaries to represent data from your surveys. A statistical summary includes such graphs as histograms, bar charts, pie charts, or box plots, as well as numbers like the range, mean, median, or mode. By organizing the data from your surveys, you were able to recommend a development plan for Murre Island.

Summary Questions

ESS surveyed a random sample of the U.S. population, analyzed the survey data, and made the following conclusions:

REPORT ON MURRE ISLAND SURVEY RESULTS: CONCLUSIONS ON TOURISM

I. About 15,000 people are expected to visit Murre Island during a 100-day summer tourist season. Each visitor is expected to stay for an average of three days. Twenty percent of the visitors will stay for one day only.

II. Accommodations must be available for 400 people to stay overnight during the peak of the season.

III. Two-thirds of people who stay overnight at similar national or state parks prefer to stay at campsites.

IV. Older people like staying in hotels and eating in restaurants.

V. Both children and adults like outdoor activities.

VI. If many different recreational activities are planned, people of all age groups will come to Murre Island.

16. a. How many people can be expected to stay at Murre Island's campground during the peak of the season?

b. Estimate how many campsites should be available.

17. Estimate how many hotel rooms will be needed.

Solutions and Samples
of student work

16. a. 267 people. Two-thirds of the 400 people who stay overnight will stay in campgrounds.

b. Answers may vary, due to different assumptions students make. Assuming 2 or 3 people share one campsite, then about 100 campsites would be enough (267 ÷ 2.5 is about 100).

17. Answers may vary. Sample response:

One-third of 400 people will stay overnight at hotels, which is about 133 people. Assuming that an average of two people share one hotel room, then 133 ÷ 2 rooms will be needed, or about 65 hotel rooms.

Hints and Comments

Materials newspapers and magazines, optional (several of each per class); scissors, optional (one pair per student); drawing paper, optional (one sheet per student); glue or tape, optional (one bottle or roll per group)

Overview Students read a report on survey results and use it to answer questions.

Planning Read and discuss the Summary with students. Discuss the different graphs that are used to represent data. Ask students what the advantages and disadvantages of each graph type are and in what situation each graph is most appropriate. Also discuss the use of range, mean, median, and mode. You may assign problems **16** and **17** as homework and/or use them as informal assessments. Students can work on these problems individually.

Comments about the Problems

16–17. Informal Assessment These problems assess students' ability to calculate with fractions, decimals, and percents. You may assign these problems as homework.

The results given in the report and the answers to problems **16** and **17** will be needed for the calculations in the exercises in the next section. They can also be used as guides for the land-use plans in the final section.

Extension Have students design a glossary of statistical terms that defines *mean, mode, median, bar chart, histogram, pie chart,* and *box plot.* You may want to ask students to look for graphs in newspapers or magazines to cut out for illustrations or to draw their own examples of these graphs.

Work Students Do

In this section, students use statistical plots to visualize patterns in data sets that are related to three environmental issues. Students use their own data and given data to help them solve open-ended questions. They make decisions that will be useful in planning a land-use proposal in Section E.

Goals

Students will:

- organize and describe data;*
- use statistical summaries and graphs;
- calculate with fractions, decimals, and percents;
- read, interpret, and use information from maps or graphical displays of data;*
- choose and use appropriate techniques to analyze data;
- make inferences based on data;*
- make decisions based on statistical information and justify these decisions.*

** These goals are addressed in this section and are assessed in other sections of this unit.*

Pacing

- approximately three 45-minute class sessions

Vocabulary

- cluster
- outlier

About the Mathematics

The mathematics in this section involves applying statistical tools and concepts to analyze and describe data. These tools and concepts can be a bar chart, a stem-and-leaf plot, a histogram, a pie chart, a stacked bar graph, a scatter plot, the mean, the median, the mode, a cluster (a set of observations that fall in one location), and an outlier (an observation that falls outside the overall pattern of the data).

This section also relates to the number strand: fractions, decimals, and percents are used to make relative comparisons. Students also find percents of percents. The given data involve some very large numbers that must be ordered or put along the axes of a graph.

Materials

- graph paper, pages 69, 73, 75, 81, 83, 87, and 89 of the Teacher Guide (seven sheets per student)
- calculators, page 85 of the Teacher Guide (one per student)

Planning Instruction

You may want to introduce this section with a discussion of environmental issues that the students are concerned about. You might ask students to create a bar chart like the one on Student Book page 27 based on a class survey.

This section deals with environmental issues related to water, energy, and waste disposal. You may choose to investigate only one of the three issues with the whole class, or the class can be divided into groups, giving each group one environmental issue to explore. The information gathered in this section can be incorporated into the land-use proposal for Murre Island which is developed in Section E.

You may wish to complete problem 1 as a whole-class activity. Students may work individually on problem 5. Students may work individually or in small groups on problems 6–8, 14–19, 30–39, and 44. They may work on the remaining problems in small groups.

Problems 15–30 are optional and may be omitted if time is a concern.

Homework

Problems 7 and 8 (page 68 of the Teacher Guide), and problems 40 and 42 (page 88 of the Teacher Guide) may be assigned as homework. The Extensions (pages 65, 71, 79, and 91 of the Teacher Guide) may be assigned as homework. The Bringing Math Home activity (page 67 of the Teacher Guide), and the Writing Opportunities (pages 73 and 91 of the Teacher Guide) may also be assigned as homework.

Planning Assessment

- Problems 6, 15, 18, and 38 can be used to informally assess students' ability to use statistical summaries and graphs.
- Problems 6, 18, and 38 can be used to informally assess students' ability to choose and use appropriate techniques to analyze data.
- Problems 17 and 35–37 can be used to informally assess students' ability to calculate with fractions, decimals, and percents.

D. DESCRIBING DATA

Environmental Problems

Although there are many advantages to bringing tourism to Murre Island, there are several disadvantages as well. For example, there may not be enough natural resources to support tourism, or a waste disposal problem may result from an increase in the number of people on the island.

ESS studied a telephone survey in which respondents were asked which environmental problems they were especially concerned about. The results of the survey are shown in the graph below.

Urgent Problems with Our Environment

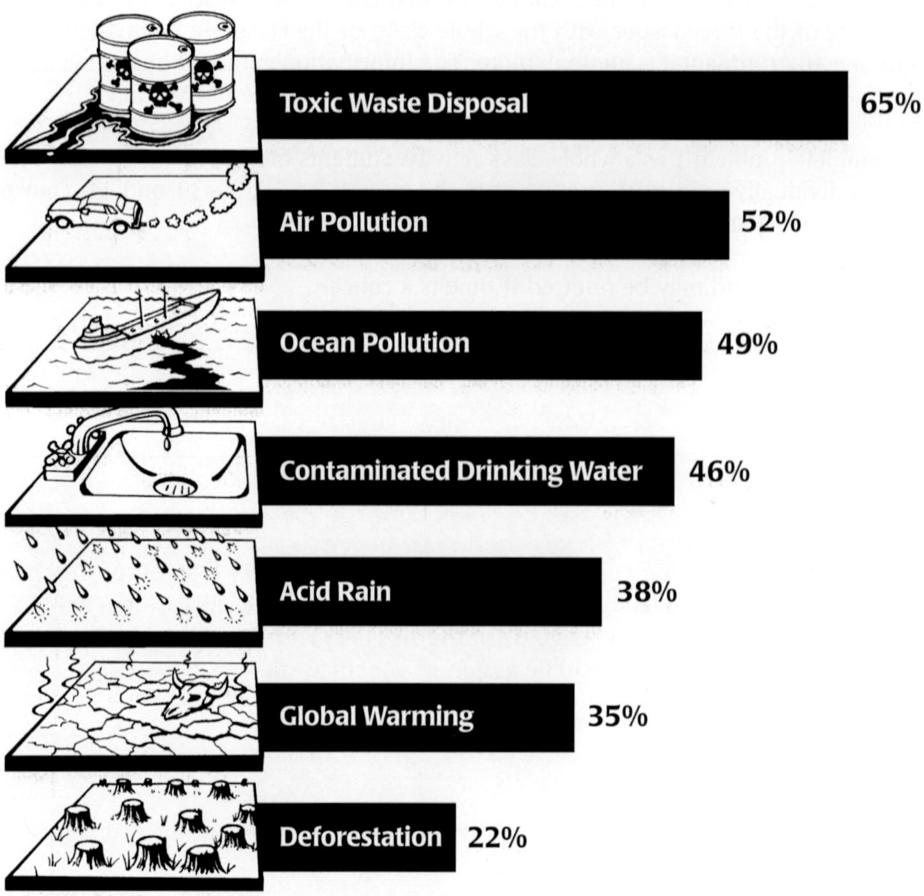

1. **a.** What are the major environmental problems identified by this sample of people?

   b. Which of these problems might affect Murre Island?

Solutions and Samples
of student work

1. **a.** toxic waste disposal, air pollution, ocean pollution, contaminated drinking water, acid rain, global warming, and deforestation

   b. Answers may vary. Some students may say ocean pollution, acid rain, and global warming.

Hints and Comments

Overview Students interpret the results of a survey about environmental problems.

Planning You may want to read the text and complete this problem as a whole-class activity.

Extension Survey the students in your class, asking them to come up with their own list of environmental problems.

Water

On Murre Island, it is possible to purify rainwater to make it suitable for drinking and other household purposes. Most of the rain that falls on Murre Island either evaporates; flows into rivers, lakes, and streams; or seeps into the soil. About 10 percent of the rain can be used for drinking and household purposes.

2. How much rainwater can be made suitable for drinking in one year on Murre Island? (*Hint:* Use the map of Murre Island on page 2 and the graph on page 3.)

3. Will there be enough water for tourists visiting Murre Island? Explain your answer.

Activity

To examine the water issue more thoroughly, you will investigate the average amount of water that a person might use daily.

Keep a record of all of the water you use during a 24-hour period. Write down what the water is used for and estimate how many gallons are used. (*Note:* You can refer to the figures on the right or ask for information from the water utility company in your community.)

Compare your water usage estimate with those of your classmates. If necessary, correct any estimates.

4. Using your estimates, determine the average amount of water that one person in your class uses in 24 hours.

Source: Data from *The World Almanac and Book of Facts 1996* (Mahwah, New Jersey: Funk and Wagnalls Corporation, 1995), 192.

Solutions and Samples
of student work

2. Estimates will vary. Sample estimate:

   47.6 billion liters, or 12.6 billion gallons per year

   The average yearly rainfall is about 1,750 millimeters or 1.750 meters per square meter.

   The total area of the island is about 272 km^2, or 272,000,000 m^2. So, the total rainfall on Murre Island (area × depth of water) is

   $272{,}000{,}000 \text{ m}^2 \times 1.750 \text{ m} = 476{,}000{,}000 \text{ m}^3$

   Therefore, the 10% of the rainfall that is not lost is 47,600,000 m^3.

   Since 1 cubic meter = 1,000 liters, there are 47,600,000,000 liters, or 12.6 billion gallons.

3. Answers and explanations will vary. Sample explanation:

   Since the island supplies over 12 billion gallons of water, there is more than enough water to supply 15,000 people staying an average of three days. Each tourist could use over 266,000 gallons per day.

4. Estimates will vary.

Hints and Comments

Overview Students investigate the amount of drinking water that is available per year on Murre Island. Then they determine the average amount of water one person uses in 24 hours.

About the Mathematics Students have had experiences with converting metric units in *Measure for Measure, Reallotment,* and *Made to Measure.* Students may remember some points of reference that are useful, for example, 1,000 millimeters = 1 meter, or 1 millimeter is $\frac{1}{1000}$ of a meter (*milli* means one–one thousandth), or 1 kilometer = 1,000 meters (*kilo* means thousand).

Planning The first part of the activity can be completed as homework. Having students keep a record of how much water they actually use will give them a better understanding of the problem. Students may work on problems **2–4** in small groups.

Comments about the Problems

2. To answer this question, students should use their answers from problem **4** on Student Book page 3 (How many millimeters per square meter of rain fall on the island?) and their answers from problem **20** on Student Book page 10 (How big is the Island?).

   You might discuss an estimate for converting liters into gallons. The number of liters divided by four is approximately the number of gallons. Similarly, the number of gallons times four is approximately the number of liters.

3. Students might answer this question after they have done the activity. An estimate will suffice. This issue will be addressed again.

4. Comparing records may be useful in explaining large differences in water usages. If large differences do exist, it may be more appropriate to use the median rather than the mean for an average, and to take into account the range of the results. Graphs can be used to picture the data.

Bringing Math Home Ask students to explain to their families how they estimated the average amount of water a person uses in 24 hours. Students can also read the water meter (or the water bill) to arrive at an estimate for their individual water use.

The table on the right provides an overview of the average amount of water used in different states for household purposes by one person during a typical day.

Domestic Water Consumption from Public Sources*

State	Per Capita Use (in gal/day)	State	Per Capita Use (in gal/day)
Alabama	95	Montana	147
Alaska	85	Nebraska	112
Arizona	145	Nevada	215
Arkansas	101	New Hampshire	99
California	133	New Jersey	75
Colorado	152	New Mexico	178
Connecticut	66	New York	93
Delaware	70	North Carolina	91
District of Columbia	279	North Dakota	77
Florida	123	Ohio	53
Georgia	117	Oklahoma	56
Hawaii	117	Oregon	127
Idaho	284	Pennsylvania	66
Illinois	86	Rhode Island	66
Indiana	115	South Carolina	75
Iowa	136	South Dakota	110
Kansas	75	Tennessee	83
Kentucky	65	Texas	143
Louisiana	145	Utah	217
Maine	115	Vermont	100
Maryland	102	Virginia	80
Massachusetts	78	Washington	146
Michigan	85	West Virginia	62
Minnesota	148	Wisconsin	54
Mississippi	73	Wyoming	185
Missouri	86	Entire U.S.	105

* Includes water for normal household purposes , such as drinking, food preparation, bathing, washing clothes and dishes, flushing toilets, and watering lawns and gardens.

Source: Data from Frits van der Leeden, Fred L. Troise, David Keith Todd, *The Water Encyclopedia*, Second Edition (Chelsea, Michigan: Lewis Publishers, Inc., 1990), 335.

5. a. Is the average water use for your class representative of your state?

b. Is the average water use for your class representative of the entire United States?

6. a. Make a plot of the information in the table.

b. Explain what the plot indicates about water usage in the United States. In your explanation, you may want to include the mean, median, or mode values and any special features of the data that stand out.

c. Why did you select the type of plot and the numbers that you used in parts **a** and **b?**

7. Why do you think that there are large differences in water usage among states?

8. a. Refer to the information about water usage on page 28. For each type of water usage, describe ways to conserve water, as well as how much water can be saved.

b. List some other ways to conserve water.

Solutions and Samples
of student work

5. **a.** Answers will vary, depending on the results of the student activity on the previous page.

   b. Answers will vary, depending on the results of the student activity on the previous page.

6. **a.** Graphs will vary. Sample stem-and-leaf plot:

Stem	Leaf
5	3 4 6
6	2 5 6 6 6
7	0 3 5 5 5 7 8
8	0 3 5 5 6 6
9	1 3 5 9
10	0 1 2
11	0 2 5 5 7 7
12	3 7
13	3 6
14	3 5 5 6 7 8
15	2
16	
17	8
18	5
19	
20	
21	5 7
22	
23	
24	
25	
26	
27	9
28	4

   b. Explanations will vary. The stem-and-leaf plot given for the solution to part **a** gives a quick overview of the range of the water consumption of the states. Most amounts are between 50 and 150 gallons.

   Outliers are Idaho (284) and the District of Columbia (279).

   The mean is 113 gallons. There is no mode, because no value is more frequent than all the others. The distribution can also be said to be bimodal (having two modes) because both 66 and 75 occur three times. The median is 100 gallons.

   c. Answers will vary, depending on students answers to parts **a** and **b.**

7. Answers will vary. Sample response:

   Some states have better conservation practices, more modern water systems, and/or more advantageous climate patterns.

8. **a.** Answers will vary. Some students may suggest taking shorter showers, buying a toilet that uses less water, shutting off the tap when brushing their teeth, and washing only full loads of clothes or dishes.

   b. Answers will vary. Possible ways to conserve water include washing the car and watering the lawn less often.

Hints and Comments

Materials graph paper (one sheet per student)

Overview Students investigate and compare the water consumption in different U.S. states. They use graphs and numbers to represent the data. They reflect on water consumption and conservation.

Planning After students have completed problem **5** individually, you may want to have students continue to work individually or in small groups on problem **6–8.** Problem **6** may be used as an informal assessment. Problems **7** and **8** may be assigned as homework.

Comments about the Problems

5. Ask students to explain any discrepancies.

5b–6. Note that the average water use of the entire United States is mentioned in the table: 105 gallons a day. If students calculate the average by adding all the other amounts and dividing by 51, they will get about 113.5 gallons per day. This is different from the U.S. total because the U.S. total is computed by adding up all the water used in the U.S. and dividing by the number of people. The average of 113.5 gives Washington, D.C., the same "weight" as California, which is not equitable, because California has many more people.

6. Informal Assessment This problem assesses students' ability to use statistical summaries and graphs and to choose appropriate techniques to analyze data.

You may have students graph data from a selection of states, for instance those in the Midwest, the East, or the South, or a random collection of around 10 states.

7–8. Homework These problems may be assigned as homework.

The data you collected on water usage in the activity on page 28 and the data in the table on page 29 provide information about household water use. When planning for Murre Island tourists, ESS must also consider campground and hotel water use.

9. Estimate the amount of water that might be used per person per day at a campground.

10. Estimate the amount of water that might be used per person per day at a hotel.

11. **a.** If you were vacationing on Murre Island, how could you conserve water?

b. Would you use less water at a hotel or a campground?

12. Estimate how much water will be needed each day on Murre Island when the hotel and campgrounds are full and people are using water as efficiently as possible.

13. Will it be necessary to bring water to Murre Island from other locations? Why or why not?

Solutions and Samples

of student work

9. Estimates will vary. Some students may estimate this by considering one shower and 4–5 uses of bathroom facilities per person per day. So, an estimate of 40–50 gallons per day per person is reasonable.

10. Estimates will vary. Students should consider bathroom facilities in each room, water use in restaurants, laundry rooms, and swimming pools. A reasonable estimate is 50–75 gallons per person per day.

11. a. Answers will vary. Sample response:

Do not allow the water to run the whole time you are brushing your teeth, and take quick showers.

b. Answers will vary. Sample response:

Campers might use more water since they require it for cooking and washing.

12. Estimates will vary. Using the answers to the problems on Student Book page 29, problems **16** and **17** on Student Book page 26, and problems **9** and **10** above, reasonable estimates would be 11,000 gallons in the campgrounds and 7,000 gallons in hotels, for a total of about 18,000 gallons per day.

13. Answers will vary. Sample response:

Using the estimate from problem **2** on Student Book page 28, it will probably be unnecessary to bring in additional water.

Hints and Comments

Overview Students use the data from the problems on the previous pages to estimate the water use of the tourists on Murre Island.

Planning Students may work on problems **9–13** in small groups. If groups of students have been assigned to different environmental problems in this section, you may ask the group assigned to the water problems to report their findings in class.

Comments about the Problems

9. The following data may be helpful in estimating water use in a campground:

- camping with central bathing and toilet facilities: 35 gallons per person per day
- overnight camping with flush toilets: 25 gallons per person per day
- trailers with bathrooms: 50 gallons per person per day.

10. The following data may be helpful in estimating water use in a hotel:

- hotel room with private bath: 30 gallons per person per day
- restaurant: 7–10 gallons per patron per day
- swimming pool: 10 gallons per swimmer per day

12. If you have not completed Section C, you should refer students to the data given on Student Book page 26.

Extension You may have groups of students make a poster about water use on Murre Island using their answers to problems **9–13.**

Energy

Energy is needed for a variety of reasons. To travel by car, for example, you need gasoline; to heat a home, you need oil, natural gas, or electricity.

The pie chart on the right shows how energy is used in the United States. As you can see, 27 percent of the energy in the United States is used for transportation.

14. Suppose that, in another country, 16 percent of the energy is used for transportation. What could be a reason for this difference between this country and the United States?

The table below on the right shows how residential energy use is distributed.

15. **a.** Make a bar chart or a pie chart to compare the data in the table.

b. Describe what your chart shows about residential energy use.

Energy Use in the United States

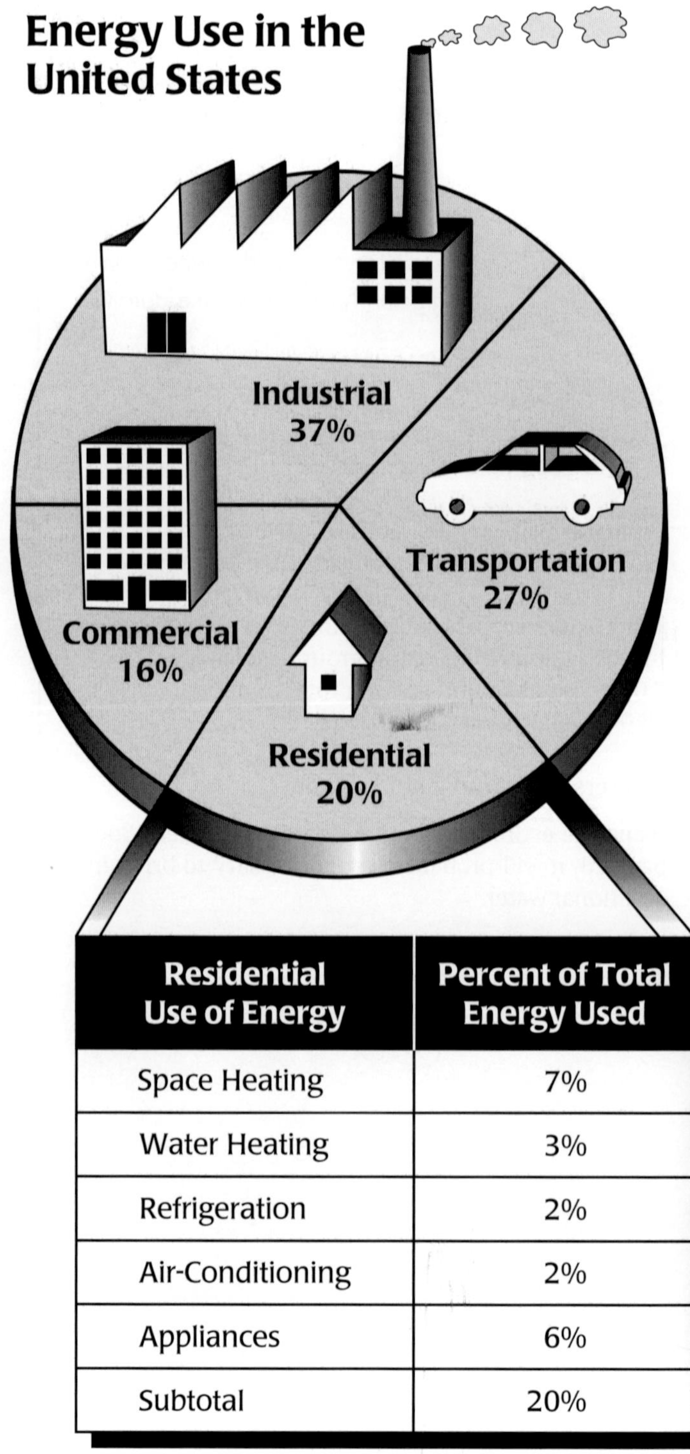

Residential Use of Energy	Percent of Total Energy Used
Space Heating	7%
Water Heating	3%
Refrigeration	2%
Air-Conditioning	2%
Appliances	6%
Subtotal	20%

Source: Data from Energy Information Administration, *State Energy Report 1993,* 1993 Residential Energy Consumption Survey.

Solutions and Samples
of student work

14. Answers will vary. Sample response:

People may travel less, or the distances are shorter, or this country has a system of transportation that is more energy efficient.

15. a. Charts will vary. Sample charts:

Residential Energy Uses

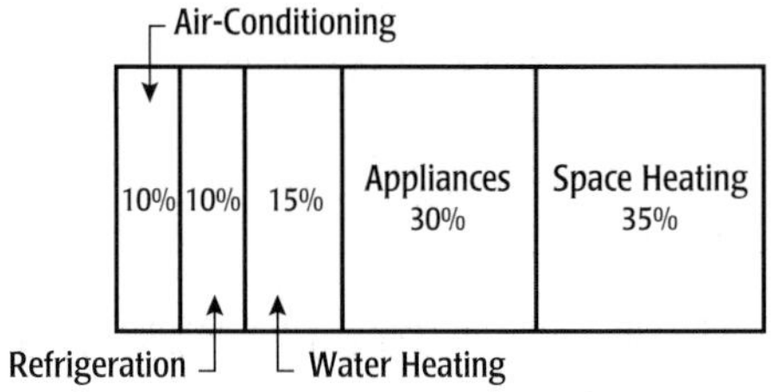

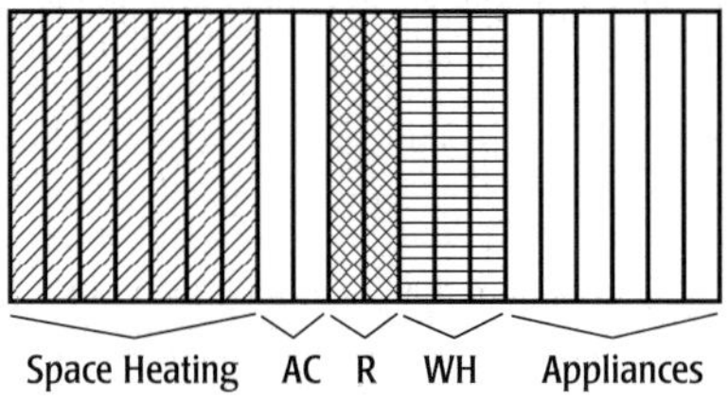

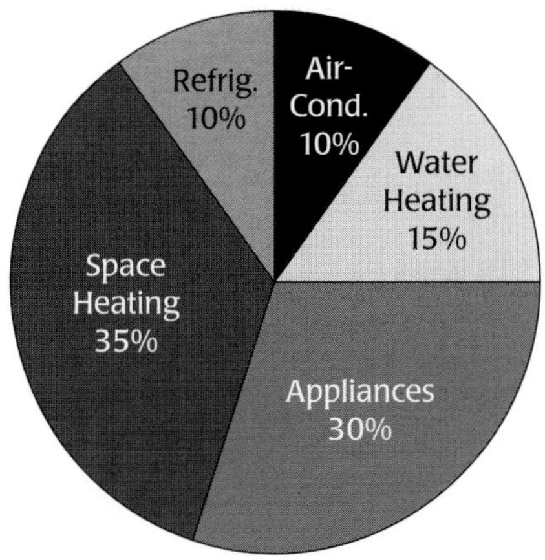

b. Descriptions will vary. Sample description:

The pie chart shows that the biggest part (about $\frac{1}{3}$) of residential energy is spent on space heating, and a bit less than $\frac{1}{3}$ is spent on appliances. Water heating uses half of the energy that appliances use. Air-conditioning and refrigeration take about $\frac{1}{10}$ of the residential energy.

Hints and Comments

Materials graph paper (one sheet per student)

Overview Students begin to investigate energy use. They make a pie chart or a bar chart to show the distribution of residential energy use.

About the Mathematics In problem **15,** students must make a pie chart of a sector of a given pie chart. This new pie chart will show the distribution within the sector. Actually, this is related to taking a percent of a percent.

Planning Students may continue working on problems **14** and **15** individually or in small groups. Problem **15** may be used as an informal assessment. If you are running out of time, problems **15–30** may be omitted without loss of continuity.

Comments about the Problems

15. Informal Assessment This problem assesses students' ability to use statistical summaries and graphs.

a. Students should either rescale the 20% to 100% or find the parts that each use takes of the total, for instance refrigerating is 2 of the 20%, which is $\frac{1}{10}$ or 10% of the pie chart.

One strategy for making a bar chart is by dividing a bar in 20 segments and coloring, respectively, 7, 3, 2, 2, and 6 of the segments. This bar can then be cut out and curled into a circle to help in making a pie chart. Students may remember this from the unit *Fraction Times.*

Writing Opportunity You may want students to write their responses to problem **15b** in their journals.

Electricity is measured in watts. However, since most electrical appliances require a lot of electrical power, a larger unit, called the kilowatt, is used. One kilowatt is equal to 1,000 watts. If something uses one kilowatt of electricity for one hour, one kilowatt-hour (kWh) of electricity is used.

The table on the right shows the most common electrical appliances used at home. Notice that home heating is not included.

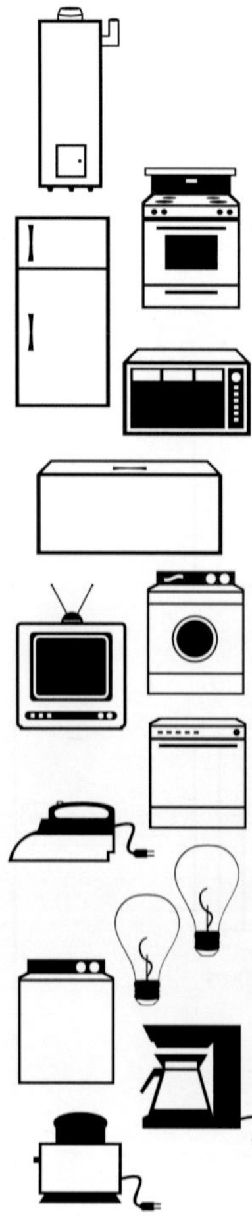

Energy Consumption of Common Electrical Appliances

Appliance	Kilowatts
Waterheater	4.5
Stove	12.2
Refrigerator	0.6
Air Conditioner	1.5
Freezer	0.34
Dryer	4.8
Color TV	0.33
Dishwasher	1.2
Iron	1.2
100-Watt Bulb	0.1
60-Watt Bulb	0.06
Washing Machine	0.5
Coffee Pot	0.9
Toaster	1.2

Source: Data from Nigel J. Hopkins, John W. Mayne, John R. Hudson, *Go Figure! The Numbers You Need for Everyday Life* (Detroit, Michigan: Visible Ink Press, 1992), 184.

16. Why do you think home heating is not included in the table?

17. How much energy do you save by using a 60-watt bulb rather than a 100-watt bulb?

18. a. Make a plot of the information in the table.

b. Explain what the plot indicates about the energy consumption of household appliances. In your explanation, you may want to include the mean, median, or mode values and any special features of the data that stand out.

c. Why did you select the type of plot and the numbers that you used in parts **a** and **b?**

Peter says, "The iron uses much more electricity than the TV, so we can save electricity by using the iron less."

19. Do you agree with Peter? Why or why not?

Solutions and Samples
of student work

16. Explanations will vary. Sample explanation:

Home heating varies between different sources such natural gas, oil, wood, and electric. This makes energy consumption for home heating more difficult to determine.

17. 40 watts, or 0.04 kilowatt

18. a. Plots will vary. Students can make a number-line plot, a box plot, a stem-and-leaf plot, or a bar graph. See sample plot below.

Stem	Leaves
0	1 1 3 3 5 6 9
1	2 2 2 5
2	
3	
4	5 8
5	
6	
7	
8	
9	
10	
11	
12	2

b. Explanations will vary. Sample explanation:

Many appliances do not require much power, less then 1 kilowatt. A stove uses considerably more energy than do the other appliances. A water heater and a dryer are also high energy users.

The mean is 29.43 ÷ 14, which is about 2.10 kilowatts. If 12.2 (the outlier) is left out, the mean is 17.23 ÷ 13, or about 1.33 kilowatts. The mode is 1.2. The median is between 0.9 and 1.2, or 1.05.

c. Answers will vary, depending on students' answers to parts **a** and **b.**

19. Answers will vary. Some students may note that the television is probably used for much longer periods of time than the iron, so it is probably using up more electricity overall.

Hints and Comments

Materials graph paper (one sheet per student)

Overview Students investigate the energy consumption of electrical appliances. They represent information in graphs.

About the Mathematics The prefix *kilo* means "thousand," so 1 kilowatt is 1,000 watts. Students should recognize this from the units kilogram and kilometer (and kilobyte).

Planning You may want students to work on problems **16–19** individually or in small groups. Problems **17** and **18** may be used as informal assessments.

Comments about the Problems

16. Utility companies may be able to provide informational brochures about electricity use.

17. Informal Assessment This problem assesses students' ability to calculate with decimals.

Students can change the amounts from kilowatts to watts to get rid of the decimal points.

18. Informal Assessment This problem assesses students' ability to use statistical summaries and graphs and to choose and use appropriate techniques to analyze data.

The large range of the data may pose a graphical challenge.

Interdisciplinary Connection Energy and electricity can be further explored in science class. Students can read the manuals or literature that come with various electrical appliances to determine their energy usage.

Activity

Determine how much electricity is used in your home each month and the electricity rates for your community. Compare your results with those of your classmates.

20. What are some possible reasons for the differences among your results?

21. Use the class data to determine the average amount of electricity used by one household each month.

Electricity is generated by burning oil, gas, and coal. These are nonrenewable, natural resources, and supplies are limited.

You can save electricity in a variety of ways.

22. List as many ways as you can think of to conserve electricity.

23. How can efficient electricity use at home affect the total amount of energy used?

One way to conserve natural resources is to make use of solar energy. Solar energy comes directly from the sun. In order to produce electricity with solar energy, solar cells are needed. Some calculators use solar cells.

24. What are some other uses of solar energy?

Solar cells on a solar-powered car capture sunlight and convert it into electricity, which, in turn, powers the motor. Built in 1995 by students from the Massachusetts Institute of Technology (MIT), the car shown on the right won Sunrayce, a biennial solar-powered car competition across America. MIT's winning entry, "Manta," reached speeds in excess of 55 miles per hour.

Source: © Warren Gretz/NREL.

Solutions and Samples
of student work

20. Explanations will vary. Sample explanations:

- The numbers of people in different households vary.
- The quantity and types of appliances present in households varies.
- Some households have electric heating.

21. Answers will vary, depending on the data collected by students.

22. Answers will vary. Sample responses:

- insulating the home
- using energy-efficient windows and doors
- using economical light bulbs
- turning the television off when no one is watching
- turning off lights when no one is in a room

23. Answers will vary. Sample response:

If you find more economical ways to heat your home, you can save a lot of energy. If you concentrate on conserving energy by turning off lights, you will not save as much.

24. Answers will vary. Sample responses:

- lighting
- water heating
- home heating

Hints and Comments

Overview Students determine the amount of electricity used monthly by a household. They reflect on electricity conservation.

Planning Students may work in small groups on the activity and on problems **20–24.**

Comments about the Problems

20–21. Let students think about which will give the best idea of average use: mean, mode, or median. A box plot or other plot may be helpful to see how the data are distributed.

Did You Know? Driving solar-powered cars is one way to be ecologically responsible. The first solar-powered car was built in 1982 by an Australian racer. In 1987, General Motors participated in (and won) the World Solar Challenge, the first transcontinental race of solar-powered vehicles. After that, General Motors teamed with Electronic Data Systems and the United States Department of Energy to create Sunrayce, a biennial intercollegiate solar-car race across America that hopes to make advanced technology a part of mainstream American life.

Sunrayce is unique because every car that competes in it is built and raced by college students. Many other solar events exist, such as the World Solar Challenge in Australia, Solar Bike Rayce USA, and the Winston Solar Challenge for high-school students.

Being Economical with Electricity

Rooftop solar energy collectors are used to heat water.

25. If you made the following changes at home, how much electricity would you save?

- Use solar energy for all water heating.
- Add insulation so that 20 percent less electricity is used for space heating.

26. What percent of the total residential energy use (see table on page 31) could be saved by making the above changes?

Energy Use on Murre Island

27. a. How do you think electricity will be used in the hotel on Murre Island?

b. Estimate the number of hours that each electrical appliance in part **a** will be used in a day.

28. Estimate the total amount of electricity (in kWh) that will be used on Murre Island each day that the hotel and the campgrounds are full.

29. List some ways in which electricity can be generated on Murre Island.

ESS
EcoSystem Savers

25. Answers will vary. Students should look at the residential percentages from Student Book page 31 and apply these to the total energy used in their homes, which they determined during the activity on Student Book page 33. Sample student response:

We use about 600 kilowatts a month. If that represents the 20% subtotal for residential energy from the pie chart on page 31, then the other amounts are as follows:

Residential Use of Energy	Percentage of Total Energy	Kilowatts Used Before Changes	Kilowatts Used After Changes
Space Heating	7%	210	168
Water Heating	3%	90	0
Refrigeration	2%	60	60
Air-Conditioning	2%	60	60
Appliances	6%	180	180
Subtotal	20%	600	468

26. 22%

27. a–b. Answers will vary. Sample response:

Appliance	Use	Hours Used Per Day
Water Heater	laundry, dishes, showers	18
Washing Machine	laundry	12
Dryer	laundry	12
Stove	restaurant	12
Refrigerator	restaurant	24
Microwave	restaurant	3
Dishwasher	restaurant	6
Other Kitchen	restaurant	3
Lights	lodge, restaurant	18
Television	guest rooms, lounge	3 per set

28. Estimates will vary. Some students may say 500 kWh per day is a reasonable estimate for the hotel, and 80 kWh per day is a reasonable estimate for 100 campsites.

29. Answers will vary. Sample response:

Electricity can be generated with the power of the wind, the sun, or ocean waves.

Overview Students determine how changes in different categories of energy affect the total energy use. Then they estimate the amount of energy that will be used on Murre Island each day.

About the Mathematics In problem **26,** the decrease in the percent of total residential energy use has to be expressed as a percent. This focuses on one of the most important aspects of percents: it is important to recognize the "whole" to which a percent is related.

Planning Students may work on problems **25–29** in small groups.

Comments about the Problems

26. Students may use their answer to problem **15a,** which shows the distribution, or they may work with the percentages that are shown in the table on page 31. If students become confused by working with percents, you may encourage them to make up an amount of energy for the total energy that was used before the changes took place, then calculate the amount that will be used after the changes, and then find the percent that will be saved.

28. Students may use the numbers mentioned at the end of Section C. Energy use at the campground will be determined by the types of facilities students decide to provide. Energy use at the hotel could be greatly reduced by using gas appliances wherever possible.

Extension You may have groups of students make a poster about electricity use on Murre Island using their answers to problems **28–30.** Then they can report their findings in class.

WASTE DISPOSAL

Do you know how much garbage is generated by one person in a week? To get an idea, you can count the number of garbage bags you fill each week at your home.

The average United States citizen throws away almost 2 kilograms of garbage each day.

30. **a.** How much waste does the average United States citizen produce in one year?

b. How much waste does everyone in the United States produce in one year?

Over the past 30 years, the total waste generated by people has increased each year. It will probably continue to increase over the next 10 years by about 2 percent per year.

31. Based on the above information, make a graph that shows how the amount of waste produced each day will increase over the next 10 years. Organize your data in a table before making your graph.

32. How will the graph change if the average person in the United States produces 1 kilogram of waste each day instead of 2?

Solutions and Samples
of student work

30. a. 730 kilograms (2 kg × 365 days)

b. Answers will vary, depending on the number of U.S. residents students come up with. Sample response:

There are 249 million people in the U.S. who each produce 730 kilograms, for a total of 181.8 billion kilograms of garbage a year.

31. Answers will vary. If students use the yearly figure they found in problem **30b,** they should divide it by 365 days. Sample response using the solution given for problem **30b,** above:

Year	0	1	2	3	4	5	6	7	8	9	10
Waste (in millions of kg)	498	508	518	528	539	550	561	572	583	595	607

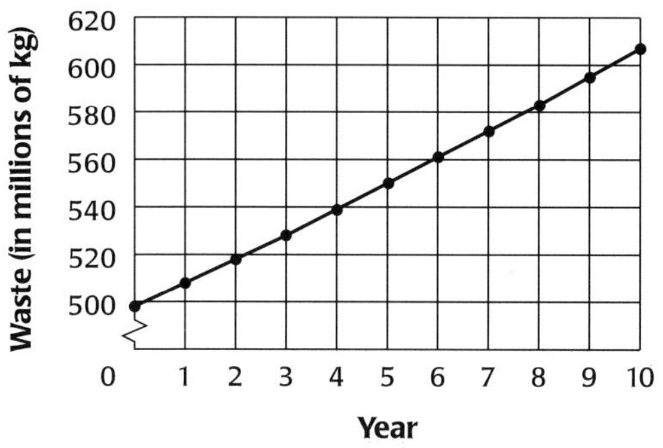

32. Every point on the graph will be lower. Every point of the graph will move to a position that is halfway between the original position and the horizontal axis.

Hints and Comments

Materials graph paper (one sheet per student)

Overview Students estimate the total production of waste in the U.S. now and through the next 10 years. They represent the waste increase in a graph.

About the Mathematics Percent increase is investigated in the unit *More or Less.* When an amount increases at regular time intervals with the same percentage, this is called exponential growth. To find the amount in the next period, you can multiply the current amount by a factor (here the factor is 1.02). Exponential growth is one of the topics in the units *Ups and Downs* and *Growth.*

Planning Problems **29–44** deal with waste disposal. Students may continue to work on problems **30–32** individually or in small groups.

Comments about the Problems

30. a. Students should know that one year has 365 days (or 366 when it is a leap year).

b. Students may not know the total number of people in the United States (according to the 1990 census, 248,709,873 people).

31. Students should start with the amount of waste produced each day. They can use for this beginning amount the production of the average United States citizen per day (2 kilograms) or the total production in the United States on one day (520 million kilograms).

Students can use any strategy to calculate the amounts in the table. Both graphs will show exponential growth, but they may look like a straight line because the time period is restricted to 10 years.

32. This is a transformation of the graph: a multiplication of the *y* values with the factor $\frac{1}{2}$. This should not be taught to the students. It is fine if they can globally describe the change of the graph. You may have students make a table of the new situation and compare this table to the table they made in problem **31** in order to check their statements.

Domestic Waste per Person per Year in The Netherlands

Year	1965	1970	1975	1980	1985	1986	1987	1988
Waste (in kg)	237	281	298	362	343	365	387	395

Source: Data from The Netherlands Central Bureau of Statistics, *Statistical Yearbook 1990 of The Netherlands* (The Netherlands: SDU Publishers, 1990), 461.

The above table shows the amount of household garbage produced per person per year in The Netherlands.

33. a. Make a graph of the data in the table.

b. How much garbage would you expect a person in The Netherlands to generate in the year 2010? Explain how you found your answer.

34. How does the amount of garbage that a person produces in one year in the United States compare with the amount that a person produces in The Netherlands?

33. a. Graphs may vary. Sample graph:

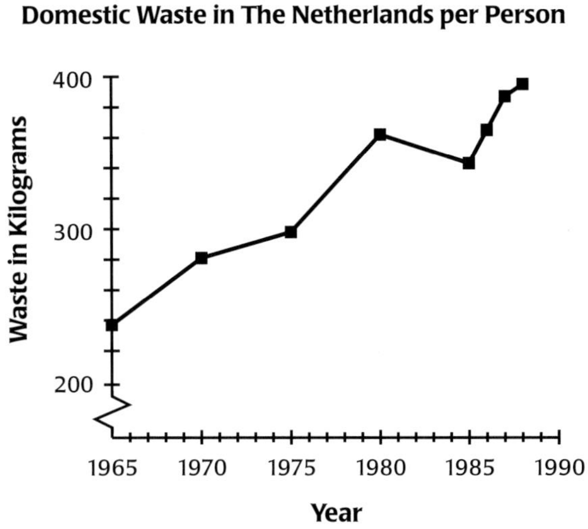

b. Answers and explanations will vary. Some students may make a projection from their graph (between 540 and 600 kg is reasonable). Another method is to use the 2% annual growth rate of garbage in the United States and the garbage level for 1988 to estimate about 610 kg for the year 2010.

34. Answers will vary. Sample response:

People in the United States produce more garbage. People in the United States produce about 730 kg of garbage per person, which is much more than a person in The Netherlands produces.

Materials graph paper (one sheet per student)

Overview Students graph data given in a table that show the amounts of waste per person in the Netherlands. They predict what the amount will be in the year 2010. Then students compare the amount of waste produced in The Netherlands with that produced in the United States.

About the Mathematics Data that are not in a table can be estimated by using the pattern of the differences in the table or by making a graph and using the shape of the graph. Finding data that are not in the table but are in the period of the given data is called *interpolation;* finding data that are outside the period of the given data is called *extrapolation.*

Planning It is important that students share and discuss their strategies for problem **33b.** Students may work on problems **33** and **34** individually or in small groups.

Comments about the Problems

33. a. A trend can be shown using a plot over time. A line graph is appropriate to use in making predictions.

b. You can predict the amount of garbage produced, but it will be an estimate based on having the same conditions as today.

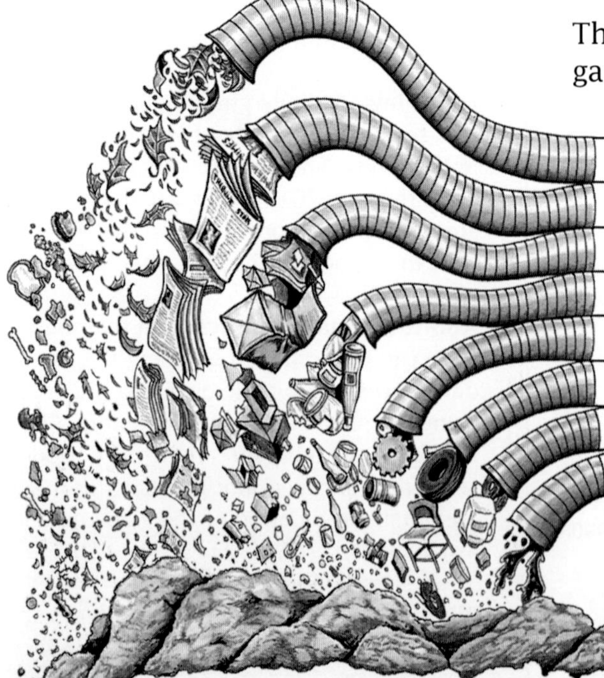

The table below shows the composition of garbage in the United States by weight.

Category	Percent
Paper and Paperboard	39%
Yard Trimmings	15%
Glass	6%
Metals	8%
Plastics	9%
Wood	7%
Food Waste	7%
Other	9%

Source: Data from Characterization of Municipal Solid Waste in the United States: 1995 Update–Executive Summary (EPA 530-S-96-001).

Getting rid of garbage is a major problem. There are three primary ways to dispose of the garbage that people produce:

- Burning (garbage incineration)
- Recycling
- Landfills

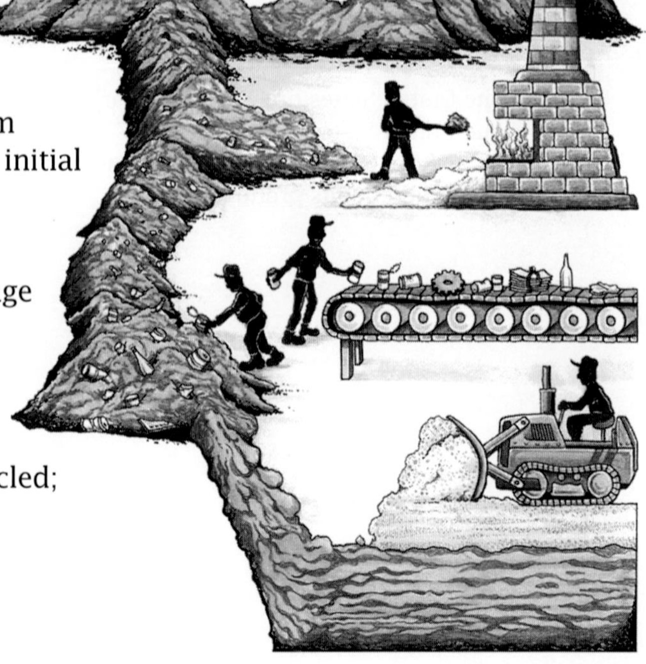

About 10 percent of the nation's garbage is disposed of by incineration. The ashes from incineration weigh about 35 percent of the initial weight of the garbage.

35. What percent of the nation's garbage remains as ashes each year after garbage incineration?

A large percent of the nation's garbage is recycled.

35% of paper and paperboard are recycled;
23% of yard trimmings are recycled;
23% of glass is recycled;
36% of metals are recycled;
5% of plastics are recycled;
10% of wood is recycled;
3% of food waste is recycled; and
11% of other garbage is recycled.

36. What percent of all garbage is recycled?

37. If the remaining garbage is brought to landfills, what percent of all garbage is brought to landfills?

Solutions and Samples
of student work

35. 3.5% (35% of 10%)

36. About 24%. The breakdown is as follows:

Paper and Paperboard:	35% of 39% is 13.65%
Yard Trimmings:	23% of 15% is 3.45%
Glass:	23% of 6% is 1.38%
Metals:	36% of 8% is 2.88%
Plastics:	5% Of 9% is 0.45%
Wood:	10% of 7% is 0.7%
Food Waste:	3% of 7% is 0.21%
Other:	11% of 9% is 1.0%

The total is 23.72%, so about 24% of the garbage (by weight) is recycled.

37. 66% or 69.5%. Sample student response:

10% is incinerated; 24% is recycled; the rest is brought to landfills, which is 100% – (10% + 24%) = 66%. If the ashes are also brought to landfills, this will add another 3.5% to the total garbage weight, making the answer 69.5%.

Hints and Comments

Materials calculators (one per student)

Overview Students investigate the composition of garbage and find the percent of garbage that is recycled and the percent that is brought to landfills.

About the Mathematics Many percent problems can be solved by using the relationship between percents and fractions. For example 10% of 7% is $\frac{1}{10}$ of 7%, which is 0.7%.

Planning You may have students work on problems **35–37** individually or in small groups. Problems **35–37** may be used as informal assessments.

Comments about the Problems

35–37. Informal Assessment These problems assess students' ability to calculate with fractions, decimals, and percents. Students must calculate a percent of a percent.

36. If students are having difficulty, you may encourage them to make estimations first. A bar can be used to give visual support. Below is a bar for the paper and paperboard category:

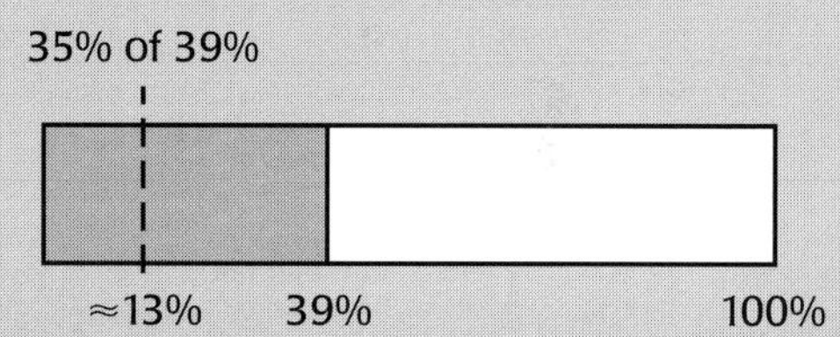

Thirty-five percent of 39% is about $\frac{1}{3}$ of 39%, which is about 13%.

Have students round the percents of the items to one decimal place. They can round their final answer, the percent of garbage that is recycled, to a whole number.

The following table shows the amount of waste generated by each state in 1991.

U.S. Solid Waste Output (1991)

State	Waste Generated (in tons per year)	State	Waste Generated (in tons per year)
Alabama[a]	4,500,000	Montana[b]	600,000
Alaska	500,000	Nebraska	1,300,000
Arizona[b]	2,900,000	Nevada	1,000,000
Arkansas[c]	2,000,000	New Hampshire[e]	1,100,000
California[a]	45,000,000	New Jersey	7,100,000
Colorado[a]	2,400,000	New Mexico[e]	1,500,000
Connecticut	2,900,000	New York	22,000,000
Delaware	750,000	North Carolina[b]	6,000,000
Florida	18,700,000	North Dakota	400,000
Georgia	4,400,000	Ohio[a]	15,700,000
Hawaii[a]	1,300,000	Oklahoma[e]	3,000,000
Idaho[b]	850,000	Oregon	3,300,000
Illinois	14,600,000	Pennsylvania	9,500,000
Indiana[c]	5,700,000	Rhode Island	1,200,000
Iowa[a]	2,300,000	South Carolina	4,000,000
Kansas	2,400,000	South Dakota[e]	800,000
Kentucky	3,500,000	Tennessee[c]	5,000,000
Louisiana	3,500,000	Texas[b]	18,000,000
Maine	950,000	Utah	1,200,000
Maryland[c]	5,100,000	Vermont	390,000
Massachusetts[b]	6,800,000	Virginia	9,000,000
Michigan[d]	11,700,000	Washington	5,100,000
Minnesota[c]	4,400,000	West Virginia	1,700,000
Mississippi	1,400,000	Wisconsin	3,400,000
Missouri[b]	7,500,000	Wyoming[b]	320,000

Data represents commercial, residential, and institutional waste, unless otherwise noted.

[a] Commercial, residential, institutional, industrial, sewage sludge, construction and demolition debris.
[b] Commercial, residential, institutional, construction and demolition.
[c] Commercial, residential, institutional, some industrial.
[d] Commercial, residential, instiitiutional, industrial.
[e] Commercial, residential, institutional, industrial, construction and demolition.

Source: Data from Jim Glenn,"The State of Garbage in America," *BioCycle Journal of Waste Recycling* (April 1992), 47.

38. a. Make a plot of the total amount of waste generated by each state.

b. Describe the information presented in your plot.

39. a. Which states appear to produce a large quantity of garbage?

b. Why might there be differences in the amounts of garbage produced by different states?

c. Make a plot that shows the relationship between the population and the amount of waste generated for each state listed on the right.

State	Population
Illinois	11,400,000
Kansas	2,500,000
New Jersey	7,700,000
Washington	4,900,000
West Virginia	1,800,000

38. a. Plots will vary. Some students may create a stem-and-leaf plot such as the one below.

0	3 3 4 5 6 7 8 8 9
1	0 1 2 2 3 3 4 5 7
2	0 3 4 4 9 9
3	0 3 4 5 5
4	0 4 4 5
5	0 1 1 7
6	0 8
7	1 5
8	
9	0 5
10	
11	7
12	
13	
14	6
15	7
16	
17	
18	0 7
19	
20	
21	
22	0
23	
24	
25	
26	
27	
28	
29	
30	
31	
32	
33	
34	
35	
36	
37	
38	
39	
40	
41	
42	
43	
44	
45	0

4 | 4 means 4.4–4.49 million tons of waste

b. Descriptions will vary, depending on the plot. The stem-and-leaf plot shows that a large cluster of states generated between 0.3 and 5.7 million tons of waste. One extreme outlier is 45 million tons of waste.

39. a. Answers may vary. Certainly, California, New York, Texas, and Florida produce large quantities of garbage.

b. Explanations will vary. Sample explanation:

States have different populations, different sources of waste, and some have more industries than others.

c.

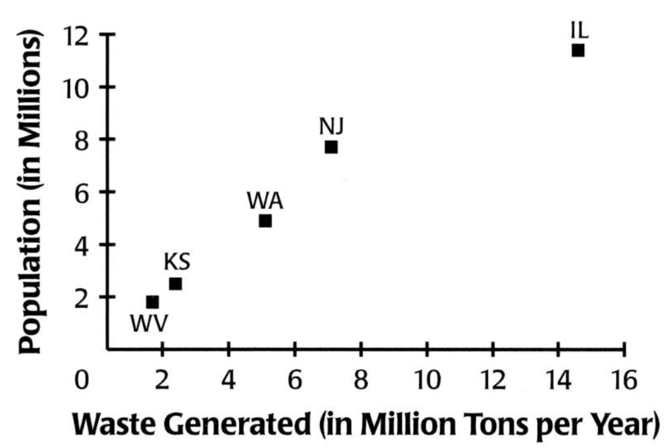

Materials graph paper (one per student)

Overview Students investigate the amounts of garbage produced in the 50 states. They represent these data in statistical graphs.

Planning You may have students work on problems **38** and **39** individually or in small groups. Problem **38** may be used as an informal assessment.

Comments about the Problems

38. Informal Assessment This problem assesses students' ability to use statistical summaries and graphs and to choose and use appropriate techniques to analyze data.

You can have students graph data from a selection of states, such as states from a specific region of the country or a random collection of about 10 states. Students might use a stem-and-leaf plot because it shows all the data points and the distribution of the points. Some students might choose to make a box plot or a histogram.

38a–39c. Students will be able to make the graph or plot most easily if they convert the data to millions of tons; 4,500,000 = 4.5 million.

Some of the states in the table on the top of page 38 have a small letter at the end of their name. The key for these small letters is provided at the end of the table.

40. a. What do the small letters represent?

b. Where does the waste recorded in the table for California come from?

c. Compare California to the other states in terms of total waste produced. How might you account for any differences?

41. The following table shows how five states managed their waste disposal in 1991.

Waste Management

State	Percent Recycled	Percent Incinerated	Percent Put in Landfill
Arkansas	5	7	88
Indiana	8	17	75
Maryland	10	17	73
Minnesota	31	25	39
Tennessee	2	9	89

Source: Data from Jim Glenn, "The State of Garbage in America," *BioCycle Journal of Waste Recycling* (April 1992), 47.

a. Make a plot or graph that allows you to compare the different methods of waste disposal for these five states.

b. Which of the states have similar waste disposal methods? Explain.

c. Which of the states brings a significantly lower percent of garbage to landfills? What might account for this?

d. The numbers for Minnesota do not total to 100 percent. What are some possible reasons for this?

42. List as many ways as you can think of to reduce the amount of garbage produced in the United States.

Solutions and Samples
of student work

40. a. The small letters represent different categories of sources of waste that are delineated in footnotes.

b. commercial, residential, institutional, industrial, sewage sludge, construction and demolition debris

c. Answers and explanations will vary. Sample response:

California produces more than twice the solid waste of any other state. It is an extreme outlier in the data set. This is probably because California has a large population and several very large cities.

41. a. Plots may vary. Sample graph:

Waste Management

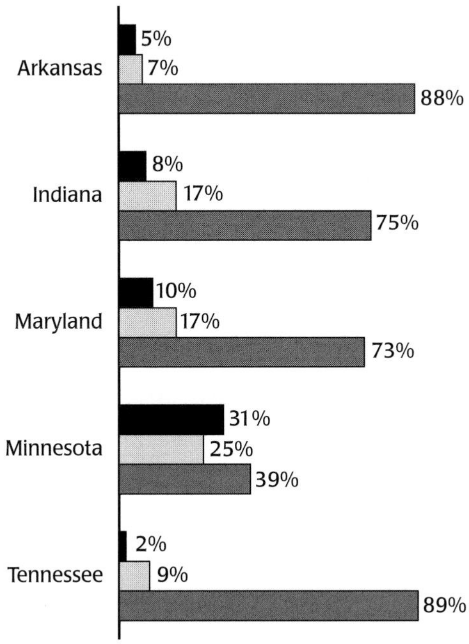

b. Indiana and Maryland are similar; the two states have nearly the same percent in each category.

c. Minnesota brings a significantly lower percent of garbage to landfills. Minnesota might have a better recycling program than the other states.

d. Answers will vary. Sample responses:

- The state may not have reported all waste.
- Some of the waste may have been shipped to out-of-state landfills.

42. Answers will vary. Sample response:

Recycle glass and plastic bottles; use paper bags rather than plastic ones; do not use Styrofoam plates and cups; and use returnable bottles.

Hints and Comments

Materials graph paper (one sheet per student)

Overview Students make a plot or graph to compare the waste management of five states.

Planning Students may work on problems **40–42** in small groups. Problems **40** and **42** may be assigned as homework.

Comments about the Problems

40, 42. Homework These problems may be assigned as homework.

41. Students can make stacked bar graphs.

Waste Disposal on Murre Island

Trash containers are needed in the campgrounds, near the hotel, and in places that tourists visit on Murre Island.

43. Estimate how much waste will be produced by tourists each season. (Remember that each visitor is expected to stay an average of three days.)

Summary

In this section, you determined how much water will be needed on Murre Island by studying water use, both in your home and per capita in each state. You also estimated the total amount of electricity that will be used on Murre Island by determining how many kilowatts a variety of appliances require. Finally, you studied how much waste a person produces, the amount of waste generated by each state, and how this waste is disposed of.

You used graphs and numbers to describe the data that you collected. You also decided which type of graph was appropriate, depending on the data.

Summary Questions

44. Pick one environmental issue that you studied in this section (water, energy, or waste) and write a paragraph describing how you used statistics to make decisions about Murre Island.

Solutions and Samples
of student work

43. Estimates will vary. Sample response:

If 15,000 visitors are expected to stay an average of three days, and each person produces 2 kilograms of waste per day, then Murre Island can expect $15{,}000 \times 3 = 45{,}000 \times 2$ kg $= 90{,}000$ kg of waste in a season.

44. Answers will vary, depending on the choice of subject. Sample student response:

I kept a record of the water I used in 24 hours (I used about 100 gallons) and then combined this amount with the amounts used by my classmates (the average was 110 gallons). These data, along with data in a table on the per capita water use in various states (most amounts were between 50 and 150 gallons), helped me to determine how much water might be used per person per day on Murre Island.

Hints and Comments

Overview Students return to Murre Island and estimate the weight of the waste that tourists will produce. They also describe how they used statistics to make decisions about Murre Island.

Planning Students may work on problem **43** in small groups. You may want to have groups of students report their findings in class. Students can continue to work on problem **44** in small groups or individually.

Extension Groups of students can make posters about waste production and management on Murre Island using their investigations and answers to problem **43.**

Writing Opportunity You may wish to have students write their responses to problem **44** in their journals.

Work Students Do

In this section, students help ESS create land-use plans for Murre Island. They design a detailed plan for one or more of the following development projects on Murre Island: a campground, hiking trails, a lodge, or areas for recreational activities. Students first read the lengthy description of each of the four development projects. This outlines essential components that must be included in the design plan, such as a map of the development area and rules and regulations that must be followed in that section of the island, as well as important points to consider, such as electricity and water requirements for the area. Students use the various maps, graphs, and tables of Murre Island from Sections A and B; the statistical summaries of the data gathered from their surveys in Section C; environmental issues involving water, energy, and waste, and corresponding data that may impact the development of Murre Island from Section D; and the mathematics from all four previous sections to help them complete the land-use plan for each development idea.

Goals

Students will:

- organize and describe data;
- use statistical summaries and graphs;
- calculate with fractions, decimals, and percents; *
- judge whether or not samples are representative and random;*
- read, interpret, and use information from maps or graphical displays of data;
- choose and use appropriate techniques to analyze data; *
- make inferences based on data;
- make decisions based on statistical information and justify the decisions.

** These goals are assessed in other sections of the unit.*

Pacing

- approximately two 45-minute class sessions.

About the Mathematics

Because *Statistics and the Environment* is a thematic unit, it uses mathematical concepts from all four strands: number, algebra, geometry, and statistics. In this section, students apply the same mathematical skills that were required in the previous sections, such as reading, interpreting, and using information from maps and graphical displays of data; making inferences based on data; and making decisions based on statistical information. Although the four development design projects are open-ended activities, students are expected to use mathematical tools, such as statistical summaries, in writing and defending their land-use plans. The mathematics in this section focuses on the higher goals of organizing information, justifying plans, and making decisions.

Materials

- Student Activity Sheets 1 and 2, pages 121 and 122 of the Teacher Guide, optional (one of each per group of students)
- drawing paper, pages 101, 103, 105, and 107 of the Teacher Guide (one or two sheets per group of students)
- students' completed survey results from Section C, pages 101, 103, 105, and 107 of the Teacher Guide
- graph paper, pages 101, 103, 105, and 107 of the Teacher Guide, optional (one sheet per group of students)
- calculators, pages 101, 103, 105, and 107 of the Teacher Guide, optional (one per student)
- centimeter rulers, pages 101, 103, 105, and 107 of the Teacher Guide, optional (one per student)

Planning Instruction

Section E is designed somewhat differently from the previous sections in this unit. It is composed of three preface pages that include a letter from ESS to Fossil Fuels, Inc., outlining the environmental restrictions on the development of Murre Island, a response letter from Fossil Fuels, and a brief introduction to the four land-use development projects. You may want to read and discuss the text on these three pages as a whole-class activity.

The remaining pages in the section provide detailed descriptions of each of the four land-use projects. You can have students work in small groups on one or more of the proposed projects. Each group then writes a proposal for the development of Murre Island with respect to their particular topic, taking into consideration environmental issues.

The final design plan for each group should then be presented to the whole class and discussed at length. At this point, you may want to integrate each of the small group proposals into one whole-class land-use plan for the island. As a culminating activity, you might have students create an attractive information brochure for tourists that outlines the class land-use plan.

Homework

Students can work on aspects of each of the four land-use development projects at home.

Planning Assessment

- Each of the four land-use development projects can be used to informally assess student's ability to organize and describe data; to use statistical summaries and graphs; and to read, interpret, and use information from maps or graphical displays of data. They also assess students' ability to make inferences based on data and to make decisions based on statistical information and justify these decisions.

E. LAND-USE PLANS FOR MURRE ISLAND

ESS now has enough information to present a detailed plan to Fossil Fuels.

ESS
EcoSystem Savers

Oliver Sax, President
Fossil Fuels, Inc.

Dear Mr. Sax:

As discussed, we are submitting a summary of our plans for Murre Island. Currently, we are considering two campgrounds, each with about 50 sites. We plan to cut down only a few trees when building these campgrounds. Each will have one rest room and several water taps at various places in the campground. We are not planning hookups for electricity at this time. One of the campgrounds will include a gift shop and a restaurant. We also intend to build a lodge for overnight visitors to the island.

As far as recreational activities are concerned, we are planning hiking trails and fishing spots. Although water sports are very popular, they will be permitted only in restricted areas along the coast. ESS will not allow water sports in any of the rivers or lakes on Murre Island because they must be protected as natural habitats.

While cars will be allowed on the island, the use and number of vehicles will be restricted. Cars will be permitted on the south side of the island; they will be prohibited on the north part of the island.

We are still debating about the number of overnight visitors to accept during the summer season.

Sincerely,

Sandy Williams

Sandy Williams, Director
ESS

Solutions and Samples

of student work

Hints and Comments

Overview Students read a letter from ESS to Fossil Fuels, Inc., which outlines ESS's land-use plans and environmental restrictions for the development of Murre Island. There are no problems for students to solve on this page.

Planning You might want to begin this section with a class discussion in which you review the activities that students have done up to this point in the unit. You may decide to read and discuss the letter as a whole-class activity or have students read it on their own. Be sure to discuss the main components of the letter with students, emphasizing ESS's responses (land-use restrictions) to environmental issues. You might ask students questions such as *What aspects of the ESS land-use plan do you agree with? disagree with? Why do you think ESS is considering only two campgrounds on the island? Why would ESS propose only one rest room for each campground and no electrical hookups at either site? Why not suggest that ESS include a gift shop and restaurant at both campgrounds and more than one lodge on the island? What is the rationale behind ESS's decisions to ban water sports and limit the number of cars on the island?*

Fossil Fuels responds to Ms. Williams' letter with the following correspondence:

FOSSIL FUELS INC.

Sandy Williams, Director
ESS

Dear Ms. Williams:

Thank you for your last letter. Your proposal includes opportunities for recreation on and preservation of Murre Island. We hope to contain the costs of building the lodge and to come up with a feasible plan for parking. We regret that you don't have any facilities for winter sports in your plans.

It is very important to us that the lodge be named after our company, but rest assured that our name will be displayed in a tasteful manner, in keeping with the natural setting of Murre Island.

We are looking forward to the final version of your plans.

Sincerely,

Oliver Sax, President
Fossil Fuels, Inc.

Solutions and Samples
of student work

Hints and Comments

Overview Students read the reply letter from Fossil Fuels, Inc., in which ESS is encouraged to finalize its plans. There are no problems for students to solve on this page.

Planning You may want to read the letter as a whole-class activity or have students read the letter on their own. Encourage students to think about ways in which to make final plans for the topics mentioned in the letter.

You might ask students questions such as *How might Fossil Fuels hold down the cost of building the lodge? What might be a feasible plan for parking at the lodge? Why do you think Fossil Fuels regrets that ESS does not have any winter sports facilities in their plans?*

More detailed plans for Murre Island must now be made. Choose one of the topics discussed on the following pages and create a plan. Each topic includes a list of things to consider and a description of what your plan must address.

You will need to look back in the unit to find various pieces of information. You may also use the results of your survey. It may be helpful to talk about your plans with other groups.

Choose one of the following topics:

- Designing a Campground
- Planning Hiking Trails
- Designing the Lodge
- Making Plans for Recreational Activities

Hints and Comments

Overview Students choose (or are assigned) one of four land-use development projects, which they will work on for the remainder of the section.

About the Mathematics In the four land-use development projects in Section E, students use statistical summaries and graphs; read, interpret, and use information from maps or graphical displays of data; make inferences based on data; and make decisions based on statistical information and justify these decisions.

The four activities involve many issues that do not deal with mathematics. However, it is important that students use mathematics to justify their choices and decisions in each of the four development projects.

Planning There are at least three ways to organize students into small groups to work on the development projects:

- You can have students choose their favorite topic and then form groups of students who have the same preference.
- You might have students first form their own groups, based on whom they would like to work with, and then have each group choose a topic (or you can assign a topic to each group).
- You can also form the student groups yourself and assign a topic to each group.

Regardless of which option you choose, be sure that each project topic is assigned to at least one group of students.

Detailed information on each of the four land-use development projects is provided on the remaining Student Book pages of this section. Be sure that students understand what is expected of them: each group must write a design plan following the given guidelines and take the considerations presented in each proposal into account. Allow students two class sessions to finish each design plan. You may then collect each group's finished product and use it as an informal assessment.

Campgrounds

CAMP MURRE

Design one of the campgrounds on Murre Island.

Your plan should include the following:

- a report on the planning process in which you present arguments for your decisions (use statistical summaries to support your arguments)
- a map of the campground to be given to campers
- campground rules and regulations
- a description of the campground to be included in a booklet of general information about Murre Island

Be sure to consider the following:

- the number of people who will stay at the campground
- the size of the campground
- the number of sites and the area of each site
- the number of trees that must be cut down
- the installation of electricity
- provisions for water
- a name for the campground
- directions to the campground
- recreational possibilities in or near the campground
- the costs of running the campground
- provisions for waste disposal
- the amount campers are charged

Solutions and Samples
of student work

Students' design plans will vary. The following sample student response provides some examples of decisions, justifications, and explanations that may be included in students' final products.

Murre Island has several unique features that make it suitable as a vacation resort. The island features miles of beautiful beaches along its coastline, which is great for swimming, sunning, or boating. It also has a inland forested mountainous area ideal for hiking. Murre Island has a variety of vegetation and rare animals. Lakes dot the island, providing peaceful settings and scenic views.

To fully enjoy Murre Island's natural surroundings, you might want to sleep under the stars in one of our two campgrounds that each feature 50 campsites. One campground, on the southern tip of the island, is easily reached by car and features a restaurant and gift shop. Our other campground, on the island's north end, is designed for the experienced campers who want to "get away from it all." This campground provides only rest room facilities and water taps. There are no electrical hookups at either campground.

The following two maps show what the campgrounds will look like.

(As part of their design plan, students should draw maps of the campgrounds that include the following:

- a map scale and legend
- main roads and nature trails
- distances to interesting places on the island
- locations of the campsites and the facilities)

Hints and Comments

Materials Student Activity Sheets 1 and 2, optional (one of each per group); drawing paper (one or two sheets per group); students' completed surveys from Section C; graph paper, optional (one sheet per group); calculators, optional (one per student); centimeter rulers, optional (one per student)

Overview Students read the guidelines and considerations for designing a campground on Murre Island. They then create a development plan that follows the guidelines and addresses the given considerations.

Planning Students may work on this development project in small groups. You may want students to complete certain portions of the project as a homework assignment. Give students two class sessions to finish the project. Collect each group's finished product if you intend to use it as an informal assessment.

Here are some criteria you might follow in evaluating the finished development plans:

- the development plan includes the topics that are to be included;
- the plan includes the topics that can be considered;
- the plan is organized and neatly done;
- all group members contributed to the final plan;
- there is evidence that students used information from previous sections;
- there is evidence that students used other sources;
- there is evidence that students correctly used the mathematics that they have learned and applied in this unit.

Comments about the Problems

The following are possible components that might be included in students' design plans for the campground:

- a description of the island
- the map of the island from Student Activity Sheet 1
- the contour map from Student Activity Sheet 2
- the letter from Sandy Williams, Director of ESS
- an estimate of the number of people camping and the number of campsites,
- students' survey results from Section C
- water, electricity, and waste disposal plans for the island

Hiking Trails

HIKE MURRE

Choose an area for the hiking trails on Murre Island. Keep in mind that various hiking trails are needed for different groups of people. Use the results of your survey and the contour maps of Murre Island to help you make your plan.

Your plan should include the following:

- a report on the planning process in which you present arguments for your decisions (use statistical summaries to support your arguments)
- a booklet with a map and a description of the trails to be given to hikers
- trail rules and regulations
- a description of the hiking trails to be included in a booklet of general information about Murre Island

Be sure to consider the following:

- scenic views and other interesting places
- the steepness of the hills
- trails for bird watchers
- water resources
- provisions for waste disposal
- the lengths of the trails and time it will take to hike them
- directions to the trails
- trail signs
- the costs of maintaining trails
- the amount hikers are charged

Solutions and Samples
of student work

Students' design plans will vary. The following sample student response provides some examples of decisions, justifications, and explanations that may be included in students' final products.

Murre Island has many scenic landscapes that feature a wide variety of trees, plants, and animals. We propose four different hiking trails, as shown on the map below. Each trail will be between 15 and 25 kilometers long and can be completed in a half day. Of course it is also possible to combine trails to make a longer trip.

The first hiking trail heads south from Bayaka along the eastern coast and then crosses the island south of Mt. Euler. It follows the western coastline, and then crosses back, ending at Bayaka.

The second hiking trail explores the northern part of the island. This primitive area features dense forests and rugged terrain that is best left to the experienced hiker.

The third hiking trail explores the relatively flat land in the southern part of the island, where an ancient tribal village is located. This is one of the easiest hiking trails to attempt.

The fourth hiking trail scales Mt. Rapul. Although it is the most difficult of the trails, the scenic view at the top makes the climb worthwhile.

(Students should draw maps of the hiking trails that include the following:

- a map scale and legend
- all hiking routes
- main roads and nature trails
- distances to interesting places on the island
- altitude indications
- side views (cross sections) of the trails)

See the solutions column on page 101 for a general description of Murre Island that students may include in their design plan for the hiking trails. Also see the Hints and Comments column for more suggestions of what students can include in their report.

Hints and Comments

Materials Student Activity Sheets 1 and 2, optional (one of each per group); drawing paper (one or two sheets per group); students' completed surveys from Section C; graph paper, optional (one sheet per group); calculators, optional (one per student); centimeter rulers, optional (one per student)

Overview Students read the guidelines and considerations for designing hiking trails on Murre Island. They then create a development plan that follows the guidelines and addresses the given considerations.

Planning Students may work on this development project in small groups. You may want students to complete certain portions of the project as a homework assignment. Give students two class sessions to finish the project. Collect each group's finished product if you intend to use it as an informal assessment. See the suggested criteria you might follow to evaluate the finished development plans in the Hints and Comments column on page 101 of the Teacher Guide.

Comments about the Problems

The following are possible components that might be included in students' design plans for the hiking trails:

- a description of the island
- the map of the island from Student Activity Sheet 1
- the contour map from Student Activity Sheet 2
- students' survey results from Section C

Lodge

MURRE LODGE

Design a lodge for Murre Island. Remember that energy must be conserved in the lodge.

Your plan should include the following:

- a report on the planning process in which you present arguments for your decisions (use statistical summaries to support your arguments)
- a floor plan of the lodge and guest rooms
- regulations and prices to be given to guests
- a description of the lodge to be included in a booklet of general information about Murre Island

Be sure to consider the following:

- the number of people to be accommodated
- the number of rooms
- access for the disabled
- economical use of water and electricity
- a name for the lodge
- landscape restoration plans
- the contents of each room
- a security plan
- directions to the lodge
- a restaurant
- provisions for waste disposal
- the costs of running the lodge

Solutions and Samples
of student work

Students' design plans will vary. The following sample student response provides some examples of decisions, justifications, and explanations that may be included in students' final products.

The lodge we propose to be built on the island will be conveniently located close to the seaport village where people will arrive by ferry. The lodge will be built entirely out of wood from the surrounding trees on the island, so not much building material needs to be shipped in from the mainland. The lodge will be well insulated to help conserve energy during colder months. Air-conditioning will be unnecessary because the lodge will be designed with large windows and high ceilings to allow the cool ocean breezes to keep temperatures comfortable during the summer.

A large container will be built on the roof to collect rainwater for various purposes. There will be a theater in which movies about the ecology of the island will be shown and in which the rangers can have meetings to tell the people about the island.

(Students should include a floor plan of the lodge and a floor plan of at least one guest room, and these floor plans should include the following aspects:

- scale and dimensions
- a legend of the symbols used
- emergency exits)

See the solutions column on page 101 for a general description of Murre Island that students may include in their design plan for the lodge. Also see the Hints and Comments column for more suggestions of what students can include in their report.

Hints and Comments

Materials Student Activity Sheets 1 and 2, optional (one of each per group); drawing paper (one or two sheets per group); students' completed surveys from Section C; graph paper, optional (one sheet per group); calculators, optional (one per student); centimeter rulers, optional (one per student)

Overview Students read the guidelines and considerations for designing a lodge on Murre Island. They then create a development plan that follows the guidelines and addresses the given considerations.

Planning Students may work on this development project in small groups. You may want students to complete certain portions of the project as a homework assignment. Give students two class sessions to finish the project. Collect each group's finished product if you intend to use it as an informal assessment. See the suggested criteria you might follow to evaluate the finished development plans in the Hints and Comments column on page 101 of the Teacher Guide.

Comments about the Problems

The following are possible components that might be included in students' design plans for the lodge:

- a description of the island
- the map of the island from Student Activity Sheet 1
- the contour map from Student Activity Sheet 2
- the letter from Sandy Williams, Director of ESS
- students' survey results from Section C
- an estimate of the number of people and an accounting of the number of rooms
- water, electricity, and waste disposal plans for the island

Recreation

In addition to hiking, several other kinds of recreational activities are possible on Murre Island. (Another group will make plans for hiking trails.)

Your plan should include the following:

- a report on the planning process in which you present arguments for your decisions (use statistical summaries to support your arguments)
- a booklet with a map showing locations of recreational activities
- a description of the recreational activities
- rules and regulations for recreational activities

Be sure to consider the following:

- guided tours
- bird and bat watching
- picnic areas
- water resources
- provisions for waste disposal
- activities in and around the lake
- activities on the beach
- directions to the various activities
- the results of your survey
- the different kinds of people who will use the facilities
- the amount tourists are charged for each activity

Solutions and Samples
of student work

Students' design plans will vary. The following sample student response provides some examples of decisions, justifications, and explanations that may be included in students' final products.

We plan to provide many recreational activities for the people who visit Murre Island. The ocean surrounding the island will be used for fishing, boating, and other water sports, such as skiing. Surfing will be popular on the west coast of the island due to the westerly winds that usually make for high waves. Only windsurfing will be allowed on the southern portion of the island, so the windsurfers will not have to worry about motorboats or surfers.

To preserve the ecology, the following rules will be enforced:

- mountain bikes may be ridden only on the bike paths; no bicycles are allowed on the nature trails or main roads.
- all cars and motorcycles may be driven only on the main roads.
- no motorboats, jet skis, or wave runners are allowed in the lakes or lagoons.

(Students should draw maps of the island that include the following:

- a map scale and legend
- main roads and nature trails
- distances to interesting places on the island
- locations of the different places where recreational activities are to be found)

See the solutions column on page 101 for a general description of Murre Island that students may include in their design plan for the hiking trails. Also see the Hints and Comments column for more suggestions of what students can include in their report.

Hints and Comments

Materials Student Activity Sheets 1 and 2, optional (one of each per group); drawing paper (one or two sheets per group); students' completed surveys from Section C; graph paper, optional (one sheet per group); calculators, optional (one per student); centimeter rulers, optional (one per student)

Overview Students read the guidelines and considerations for making a plan for additional recreational activities on Murre Island. They then create a development plan that follows the guidelines and addresses the given considerations.

Planning Students may work on this development project in small groups. You may want students to complete certain portions of the project as a homework assignment. Give students two class sessions to finish the project. Collect each group's finished product if you intend to use it as an informal assessment. See the suggested criteria you might follow to evaluate the finished development plans in the Hints and Comments column on page 101 of the Teacher Guide.

Comments about the Problems

The following are possible components that might be included in students' design plans for additional recreational activities:

- a description of the island
- the letter from Sandy Williams, Director of ESS
- the map of the island from Student Activity Sheet 1
- the contour map from Student Activity Sheet 2
- students' survey results from Section C

Assessment Overview

Students work on three assessment activities that you can use to determine what each student knows about reading, interpreting, and using data from maps, graphs, and statistical summaries and their ability to make and justify decisions based on statistical information.

Goals	Assessment Opportunities
• organize and describe data	Garbage Lake Superior Island
• use statistical summaries and graphs	Garbage
• calculate with fractions, decimals, and percents	Garbage
• judge whether or not samples are representative and random	Lake Superior Island
• read, interpret, and use information from maps or graphical displays of data	Hiking on an Island
• choose and use appropriate techniques to analyze data	Garbage
• make inferences based on data	Garbage Lake Superior Island
• make decisions based on statistical information and justify these decisions	Garbage

Pacing

When combined, the three assessment activities will take approximately two 45-minute class sessions. See the Planning Assessment section for further suggestions as to how you might use these assessment activities.

About the Mathematics

The three assessment activities evaluate most of the major goals of the *Statistics and the Environment* unit. Refer to the Goals and Assessment Opportunities sections above for information regarding the specific goals assessed in each assessment activity. Students may use different strategies to solve each problem. Their choice of strategies may indicate their level of comprehension of the problem. Consider how well students' strategies address the problem, as well as how successful students are at applying their strategies in the problem-solving process.

The Garbage assessment focuses on reading and interpreting a table, representing data with a graph, and interpreting and writing statements, while the Lake Superior Island assessment focuses on reading and interpreting information and sampling. The Hiking on an Island assessment focuses on contour maps and profiles.

Materials

- assessments, pages 130–133 of the Teacher Guide (one of each per student)
- graph paper or Student Activity Sheet 5, pages 111 and 113 of the Teacher Guide, optional (one per student)
- rulers, pages 111 and 113 of the Teacher Guide, optional (one per student)
- calculators, pages 111 and 113 of the Teacher Guide, optional (one per student)
- world atlases, page 117 of the Teacher Guide, optional (one per group)

Planning Instruction

You may want students to work on these assessment activities individually so that you can evaluate each student's understanding and abilities. Make sure that you allow enough time for students to complete the assessment activities. Students are free to solve each problem in their own way. They may choose to use any of the models introduced and developed in this unit to solve problems that do not call for a specific model.

You might use a combination of two or more of the following activities as end-of-unit assessments:

- Problems from Section C, in which students must show an understanding of summarizing and representing data.
- The four projects in Section E, in which students design a plan using statistical data presented earlier in the unit.
- The three end-of-unit assessment activities provided in this section.

Scoring

Answers are provided for all assessment tasks. The method of scoring the problems depends on the types of questions in the assessment. Most questions require students to explain their reasoning or justify their answers. For these questions, the reasoning used by student in solving the problems, as well as the correctness of the answers, should be considered as part of your grading scheme. A holistic scoring plan may be used to evaluate an entire task. For example, after reviewing a student's work, you may assign a key word, such as *emerging, developing, accomplishing,* or *exceeding,* to describe their mathematical problem-solving, reasoning, and communication.

On other tasks, it may be more appropriate to assign point values for each response. Students' progress towards the goals of the unit should also be considered. Descriptive statements that include details of a student's solution to an assessment activity may be recorded. These statements could provide insight into a student's progress toward a specific goal of the unit. Descriptive statements are often more informative than simply test scores, and can be used to document student growth in mathematics over time.

GARBAGE

Use additional paper as needed.

Below is a table that shows the amount of garbage per person produced in 11 different countries. Note that most countries produce less garbage than the United States.

Garbage per Person in Several Countries in 1985

Country	% Paper	% Plastics	% Glass	% Metals	% Misc.	Total (in kilograms)
Australia	26.0	6.1	15.1	7.0	45.8	680
Canada	36.5	4.7	6.6	6.6	45.7	640
France	27.5	4.5	7.5	6.5	54.0	270
Germany	17.9	5.4	9.2	3.2	64.3	320
Italy	22.3	7.2	6.2	3.1	61.6	260
Japan	38.3	7.7	1.3	1.4	51.3	340
Luxembourg	17.2	6.4	7.2	2.6	66.6	360
Portugal	19.0	3.0	3.0	3.5	71.5	220
Spain	15.0	6.0	6.0	2.5	70.5	280
The Netherlands	22.8	6.8	7.2	3.4	59.8	450
United States	34.7	6.7	9.0	8.8	40.8	740

Source: Data from The Netherlands Central Bureau of Statitistics, *Statistical Yearbook 1990 of The Netherlands* (The Netherlands: SDU Publishers, 1990), 463

1. Which country produced the most garbage per person? Which country produced the least garbage per person?

2. Use the data in the last column of the table to make a graph. You can make a pie chart, a bar graph, or a stem-and-leaf-plot. Explain your choice.

3. Which countries in the table seem to produce a lot of garbage? How is this illustrated in your graph?

Solutions and Samples
of student work

1. The United States produced most garbage per person. Portugal produced the least amount of garbage per person.

2. Graphs and explanations will vary. Sample graphs and explanations:

Stem	Leaf
2	2 6 7 8
3	2 4 6
4	5
5	
6	4 8
7	4

Key: 4 | 5 = 450 Kilograms

I chose a stem-and-leaf-plot because it clearly shows any outliers. According to the plot, there are three outliers. They are Canada, Australia, and the United States.

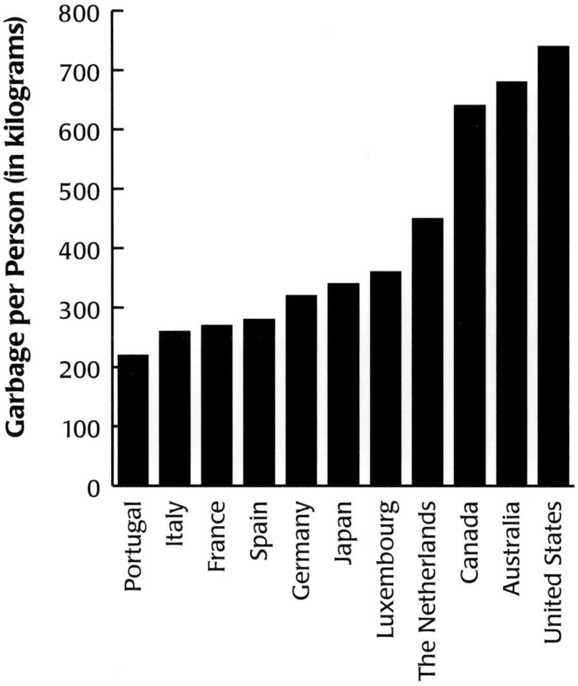

I chose a bar graph because it is an easy way to compare the garbage produced per person in the 11 countries.

3. Australia, Canada and the United States produce a lot of garbage. In the stem-and-leaf plot, these countries are grouped (clustered) together at the bottom of the plot, and in the bar graph, these countries have the highest bars.

Hints and Comments

Materials Garbage assessment, pages 130 and 131 of the Teacher Guide (one of each per student); graph paper or Student Activity Sheet 5, optional (one per student); rulers, optional (one per student); calculators, optional (one per student)

Overview Students investigate a table that shows the amount of garbage produced per person for 11 countries. They then make a graph to display the given data.

About the Mathematics These assessment activities evaluate students' ability to organize and describe data; to use statistical summaries and graphs; to calculate with fractions, decimals, and percents; to read, interpret, and use information from maps or graphical displays of data; to make inferences and decisions based on statistical information and justify these decisions; and to choose and use appropriate techniques to analyze data.

Planning You may want students to work on these assessment activities individually.

Comments about the Problems

1. The intent of this question is to have students read and interpret the data in the table.

2. A student's explanation for his or her graph choice is as important as the graph itself. Give full credit to students who provide a good explanation for their graph choice, even though the type of graph chosen may not be as appropriate as other graphical representations.

3. Some students may make an absolute comparison by examining only the data in the table. Other students may make a relative comparison by comparing the table data in relation to the population of each country. Such students should receive extra credit. If the population of each country is taken into account, Australia and Canada do not have as much garbage as other countries.

Students' responses should indicate that there is a cluster of countries that produce a large amount of garbage per person.

GARBAGE

Use additional paper as needed.

Not all garbage is incinerated or dumped. Paper, plastics, glass, and metals can be recycled. Bill, a spokesman for the Environmental Society, looks at the table on the first page of this assessment and says:

> "Only 40.8% of the garbage in the United States is not recycled.
>
> The rest (paper, plastics, glass, and metal) is recycled.
>
> In all other countries, a lower percentage of garbage is recycled, so the United States produces the least amount of garbage per person!"

4. Do you agree with the above statement? Why or why not?

5. Look carefully at the data in the table, and write two statements or two problems using the data. Explain how you came up with your statements or give the answers to the problems.

Solutions and Samples
of student work

4. Answers will vary. Students should note that percentagewise the U.S. indeed produces a small amount of non-recycled garbage. However, this is still 40% of 740 kg = 296 kg, which is more than the total amount of garbage produced per person in Portugal, Spain, Italy, and France.

5. Answers will vary. Sample responses:
   - The Netherlands and Italy have a similar composition of garbage.
   - Portugal and the United States differ a lot in the composition of garbage.

Hints and Comments

Materials Garbage assessment, pages 130 and 131 of the Teacher Guide (one of each per student); graph paper or Student Activity Sheet 5, optional (one per student); rulers, optional (one per student); calculators, optional (one per student)

Overview Students interpret and analyze the given data. They then write two statements or problems regarding the table data.

Planning Students may continue to work on these assessment activities individually.

Comments about the Problems

4. Students' explanations are more important than whether or not they agree with the given statement.
5. After students initially write their two statements or problems, you may want to assign the Extension below as an additional assessment problem.

Extension The table allows for many more statements and problems. You may want to ask students to write a few paragraphs that expand upon their response to problem **5.**

LAKE SUPERIOR ISLAND

Use additional paper as needed.

The following journal article is about the commercial development of an island along the southern shore of Lake Superior. Carefully read the journal article.

Lake Superior Island to Be Developed

Grand Island, the largest island along the southern shore of Lake Superior, will be developed commercially as a national recreation area. The U.S. Forest Service has developed a plan for the 13,500-acre island that will include a lodge, some roads, hunting, and numerous recreational activities, but it will also leave large areas of the island undeveloped in a near-natural state. The recreational area should draw people from as far away as Madison, Wisconsin, because there are no other such recreational areas in the state of Wisconsin.

The plan is a compromise between the interests of local businesses and environmentalists. Even so, some environmentalists are still angered by the plan because they fear that the upcoming development will ruin the natural beauty of the island, which includes pristine white beaches, lush hardwood forests, and towering sandstone cliffs. They also worry about the wildlife. The 4-by-7-mile island is currently inhabited by bears, otters, coyotes, martens, foxes, eagles, hawks, grouse, songbirds, and deer.

The recreational plan includes seasonal hunting, trapping, and fishing. Campgrounds will be constructed along with hiking and biking trails. During the summer, there will be extended trails for hikers and mountain bikers. Rock climbing and jet-skiing are the only activities that will be banned at Grand Island.

1. a. List the concerns of the environmentalists who are against the development plan. Do you think these concerns are valid? Explain.

b. What other environmental issues can you think of that should be considered before the island is developed as a vacation spot?

The island developers want to know how many people to expect, and they want to know what recreational activities people are most interested in. The developers think that most people will come on a day trip only. The developers design a survey.

2. a. Explain why each of the following samples should or should not be used by the developers to complete their survey.

- a representative sample from the United States
- a representative sample from Madison, Wisconsin
- a representative sample from the states Michigan, Wisconsin, and Minnesota

b. Suggest another representative sample that the developers might use.

Solutions and Samples
of student work

1. **a.** Answers will vary. Sample response:

   The concerns listed in the article are that the development plan would ruin the island—its beaches, hardwood forests, and sandstone cliffs. Depending on how large of an area will be left undeveloped and where the undeveloped area will be located, these are valid concerns.

   b. Answers will vary. Sample responses:

   - How much will the automobiles on the new roads pollute the air on Grand Island?
   - Would the land development destroy the natural habitats of the animals or endanger certain plants?

2. **a.** Answers will vary. Sample responses:

   - A representative sample from the United States is probably too large, since you cannot expect that people from the whole U.S. will come to this island.
   - A representative sample from Madison may be too small. Madison is not a very large city, and although many nature lovers live in Madison, you can expect that people from other (larger) cities such as Minneapolis or Milwaukee.
   - A sample from Michigan, Wisconsin, and Minnesota is probably best, since these states are located along Lake Superior.

   b. Answers will vary. Some students may suggest more specific samples, such as only nature lovers from the neighboring states or from Canada.

Hints and Comments

Materials Lake Superior Island assessment, page 132 of the Teacher Guide (one per student)

Overview Students read a journal article about the commercial development of an island. They list the main arguments of environmentalists against the project. Students also recommend an appropriate population sample for a survey to gather more information about what recreational activities people would be most interested in doing on the island.

About the Mathematics These assessment activities evaluate students' ability to organize, describe, and make inferences from data. They also assess their ability to judge whether or not samples are representative or random. Students should be able to show an understanding of representative sampling.

Planning If students are having difficulty reading the article, you may want to read it aloud to the class. You may assign this assessment as homework.

HIKING ON AN ISLAND

Use additional paper as needed.

Below you see a contour map and a drawing of a small island. The numbers in the contour map indicate altitudes in meters.

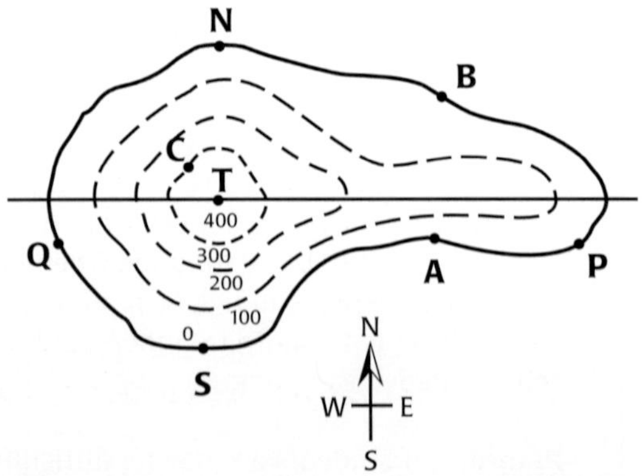

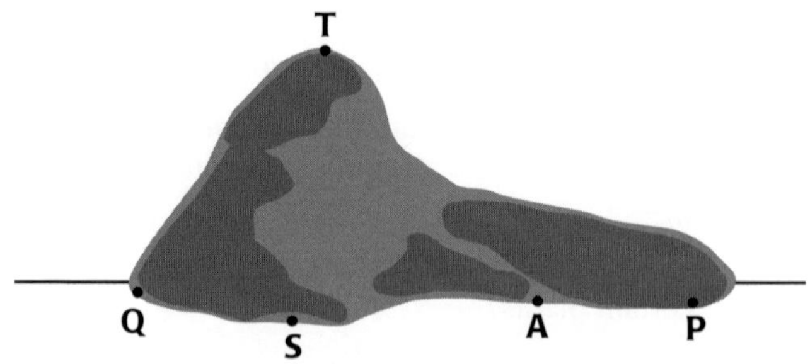

1. Draw a profile to show the varying altitudes on a hiking route that begins at the most northern point, *N*, and goes straight to the most southern point, *S*.

On the right is a profile of one hiking route on the island.

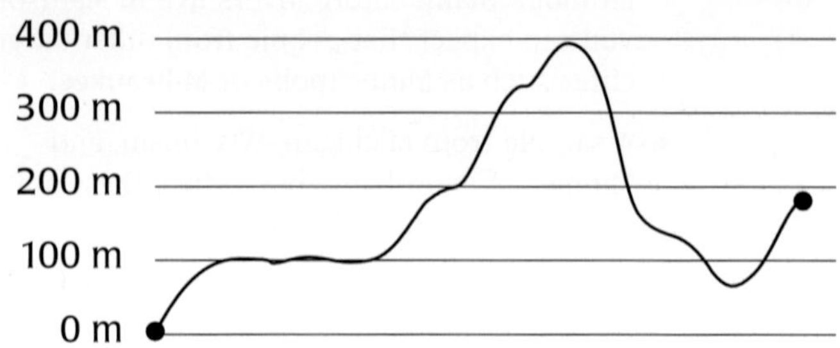

2. Describe how this route ascends and descends.

3. On the above contour map, indicate a route that matches this profile (draw in the route with your pencil).

The following are descriptions of two hiking routes from point B to point Q:

I. *From point B, cross the island, walking straight to point A.*
From point A, travel along the coastline in a westerly direction till you reach point Q.

II. *From point B, walk straight to point C, from which you can get a beautiful view of the island. From point C, walk straight to point Q.*

4. Which of the two hiking routes do you prefer. Explain.

Solutions and Samples
of student work

1.

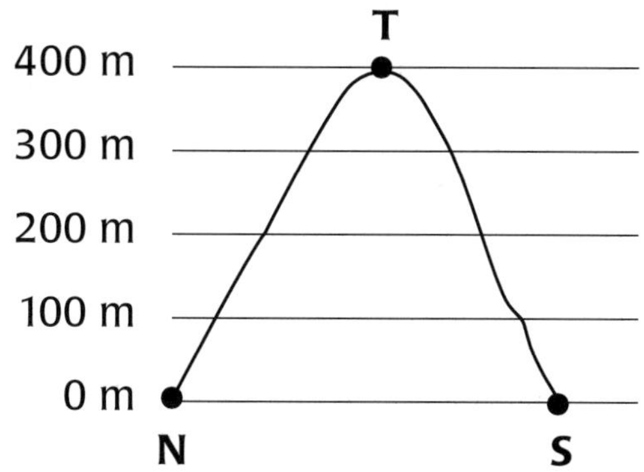

2. Descriptions may vary. Sample description:

   The route first ascends 400 meters. Then it descends about 350 meters, and ascends again 100 meters, finishing at an elevation of 150 meters.

3. Several routes are possible.

   Two possibilities are:

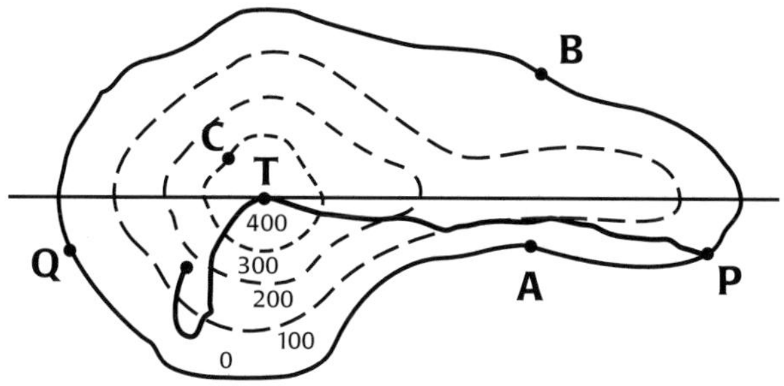

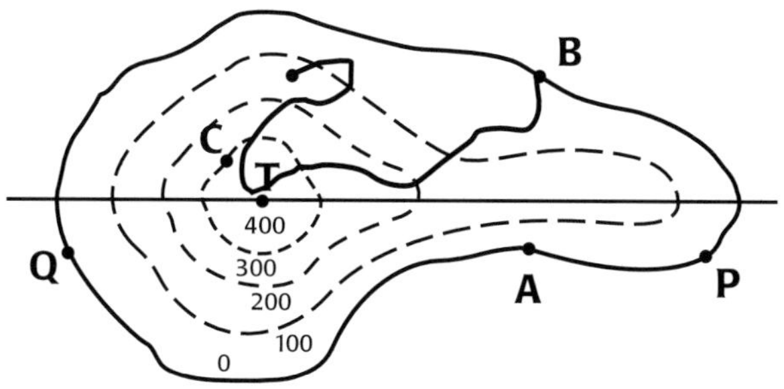

4. Answers will vary. Sample responses:

   Route I is longer but is not as steep, so it may be easier to hike. This is a route along the shore, and I would prefer this route and view.

   Route II is shorter and also quite steep. I would like the challenging climb and would prefer the view from a higher altitude.

Hints and Comments

Materials Hiking on an Island assessment, page 133 of the Teacher Guide (one per student); world atlases, optional (one per group)

Overview Students explore an island's contour map to investigate several hiking routes. They interpret descriptions of given routes and make and draw profiles of other hiking routes.

About the Mathematics These assessment activities evaluate students' ability to read, interpret, and use information from maps. Students also need to justify the choices they make.

Planning This assessment can be used anytime after students finish Section A or used as an end-of-unit assessment. Students may work on these assessment activities individually.

Comments about the Problems

1. You might suggest that students draw a line from *N* to *S* and ask them to draw the profile that corresponds with this line (that represents a hiking route from *N* to *S* straight over the top *T*).

3. There are many possible routes. The following criteria should be used in evaluating students' routes:
   - the route should start at the coastline;
   - the route should continue for a distance at an altitude of about 100 meters;
   - it should extend to the top of the island and back;
   - it should contain a designated point at which a person goes from traveling uphill to downhill;
   - it should end at about 150 meters.

   Be lenient in evaluating students' routes.

4. Students' explanations are more important than their choices of hiking routes.

Bringing Math Home As an extension to this assessment, you may ask students to look up a contour map in an atlas, and describe a route on this map.

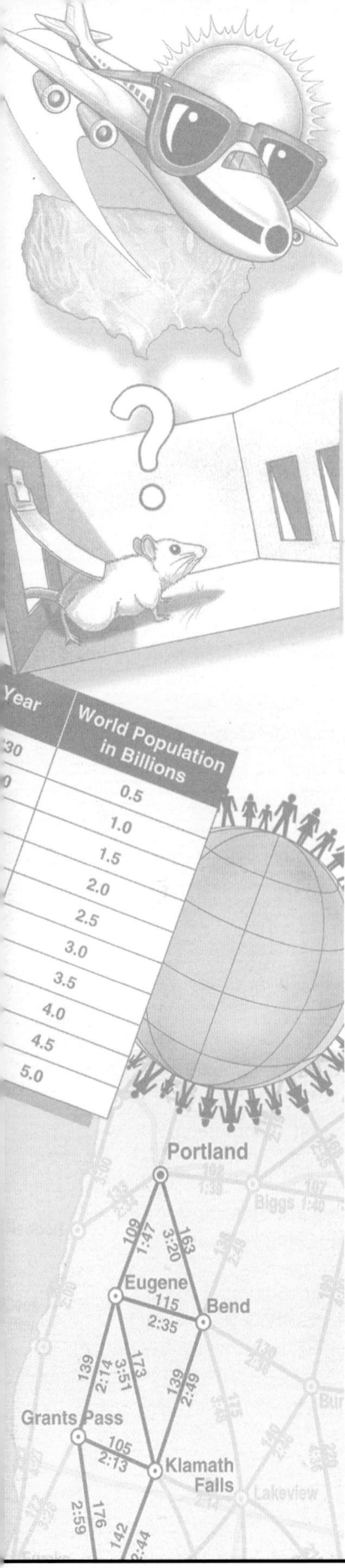

Statistics and the Environment

Glossary

The Glossary defines all vocabulary words listed on the Section Opener pages. It includes the mathematical terms that may be new to students, as well as words having to do with the contexts introduced in the unit. (*Note:* The Student Book does not include a glossary in order for students to construct their own definitions, based on their personal experiences with the unit activities.)

The definitions below are specific to the use of the terms in this unit. The page numbers given are from this Teacher Guide.

cluster (p. 62) a group of points that are close to each other on a plot

contour map (p. 16) a map having lines connecting the points that have the same elevation

outlier (p. 62) a point that lies outside the cluster in a scatter plot

population (p. 28) any group of individuals

profile (p. 18) a side view, or outline, of a shape

random sample (p. 52) a sample chosen from a population in such a way that everyone in the population has an equal chance of being selected for the sample

representative sample (p. 28) a sample that represents the population that it is taken from; in other words, a small group that has the same characteristics as a larger group

sample (p. 28) a group taken from a larger population

statistical summary (p. 50) a summary of data in the form of a graph (such as a histogram, bar chart, pie chart, or box plot) or numbers (such as the range, mean, median, or mode)

BRITANNICA Mathematics in Context

Blackline Masters

Letter to the Family

Dear Family,

Very soon your child will begin the *Mathematics in Context* unit *Statistics and the Environment.* Below is a letter to your child that opens the unit, describing the unit and its goals.

This unit focuses on describing, displaying, and interpreting data. As your child works on the unit, you may want to help him or her look for graphs in newspapers and magazines. Have your child examine the graphs to see whether there are any patterns in the data. Does some aspect of the data surprise you or your child? Can your child explain why the data looks the way it does? Is the graph well organized? Does your child understand what the graph displays? Is any information missing?

You and your child might also look for the results of a survey in a newspaper or magazine. Who was surveyed, and who responded? Were the people surveyed chosen at random? What decisions might be made based on the results of the survey? Ask your child to think of reasons that the survey might have been conducted. Were the survey results used to support an argument?

We hope you enjoy helping your child explore the many important uses of statistics.

Sincerely,

The Mathematics in Context Development Team

Dear Student,

Welcome to *Statistics and the Environment.*

In this unit, you will develop a plan for turning Murre Island, a fictitious island, into a nature reserve.

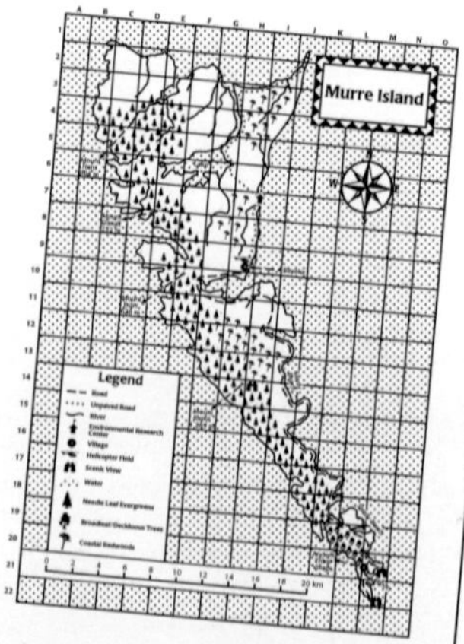

To develop an environmentally responsible plan, you will study the island's climate, topography, forests, and wildlife by organizing, displaying, and interpreting data. For example, you will study data on old-growth forests and national parks in order to make decisions about how to preserve old-growth forests on Murre Island.

You will also design and conduct a survey to determine how people like to spend their vacations. You will then tally the results of your surveys and make graphs to display the results. Based on these results, you will decide which facilities on Murre Island will attract visitors.

Finally, you will determine the water, energy, and waste disposal requirements for Murre Island. You will study the water and energy usage and waste production of both your household and of each state in the United States in order to make related decisions about Murre Island.

We hope that you enjoy using statistics to make informed and environmentally responsible decisions.

Sincerely,

The Mathematics in Context Development Team

Name ______________________________

Student Activity Sheet 1

Use with *Statistics and the Environment,* pages 2, 6, and 15.

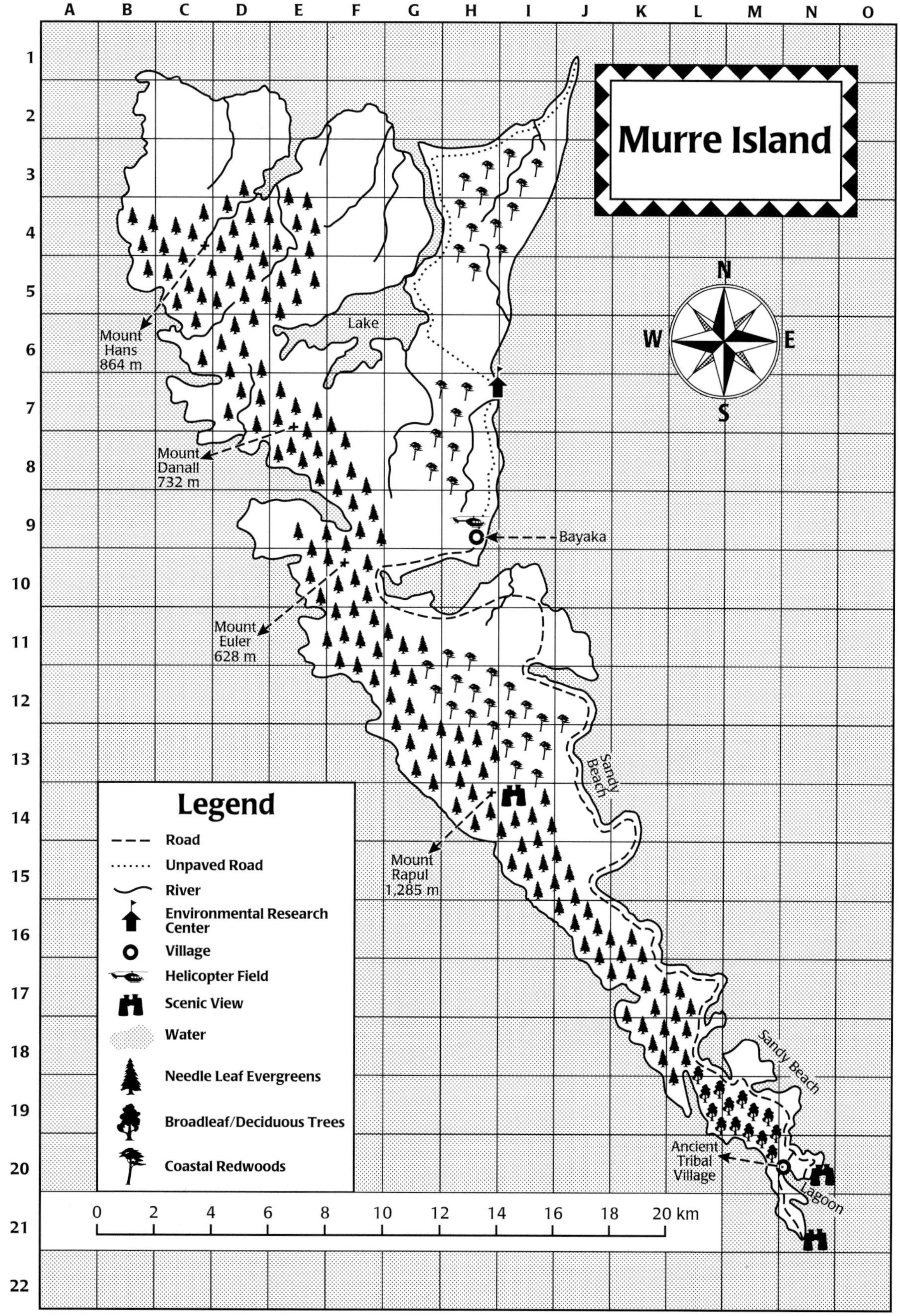

Student Activity Sheet 2

Name ______________________________

Use with *Statistics and the Environment*, page 6.

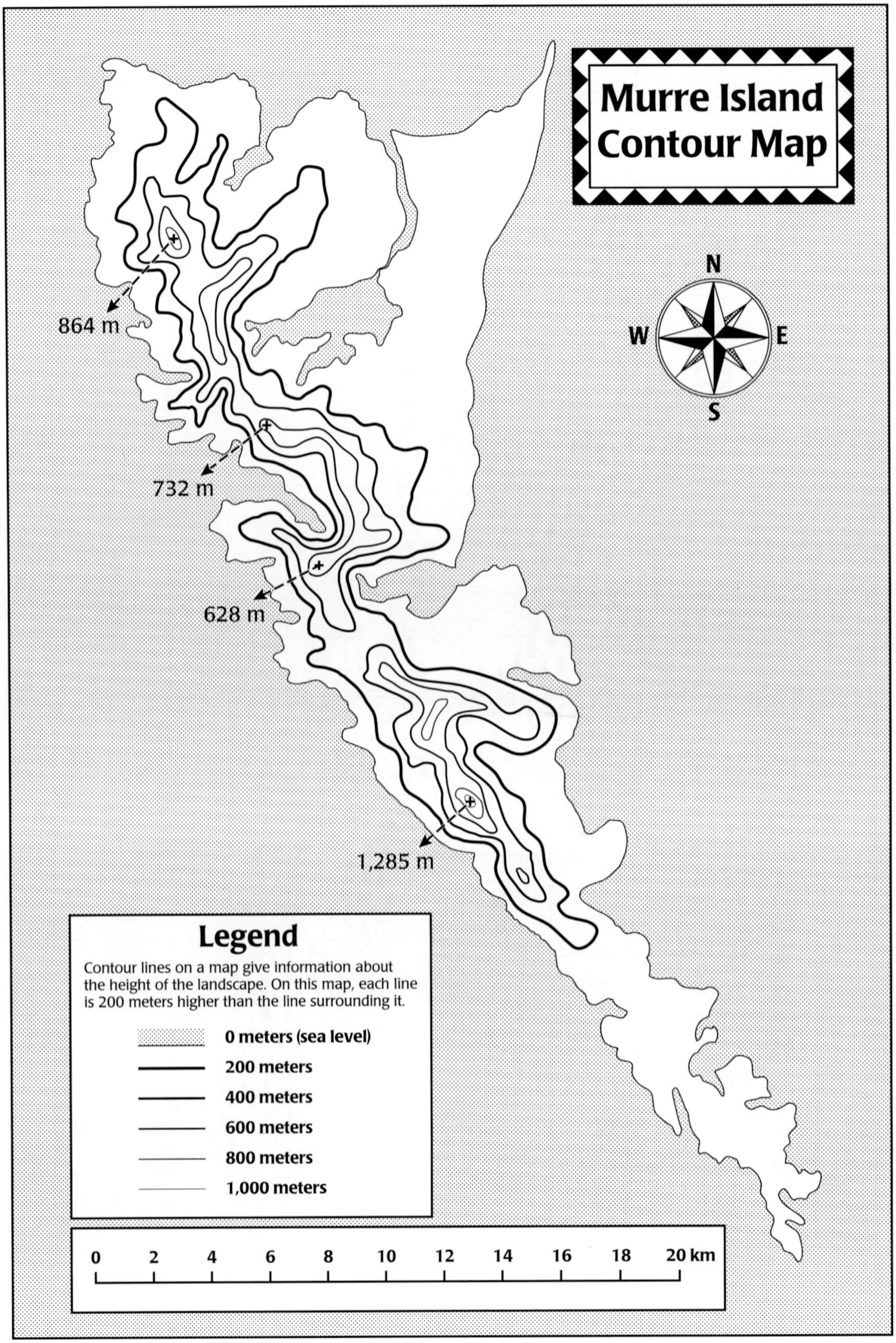

Name ____________________________________

Student Activity Sheet 3

Use with *Statistics and the Environment*, page 7.

14. c. Draw a profile of trail A from the start to the summit.

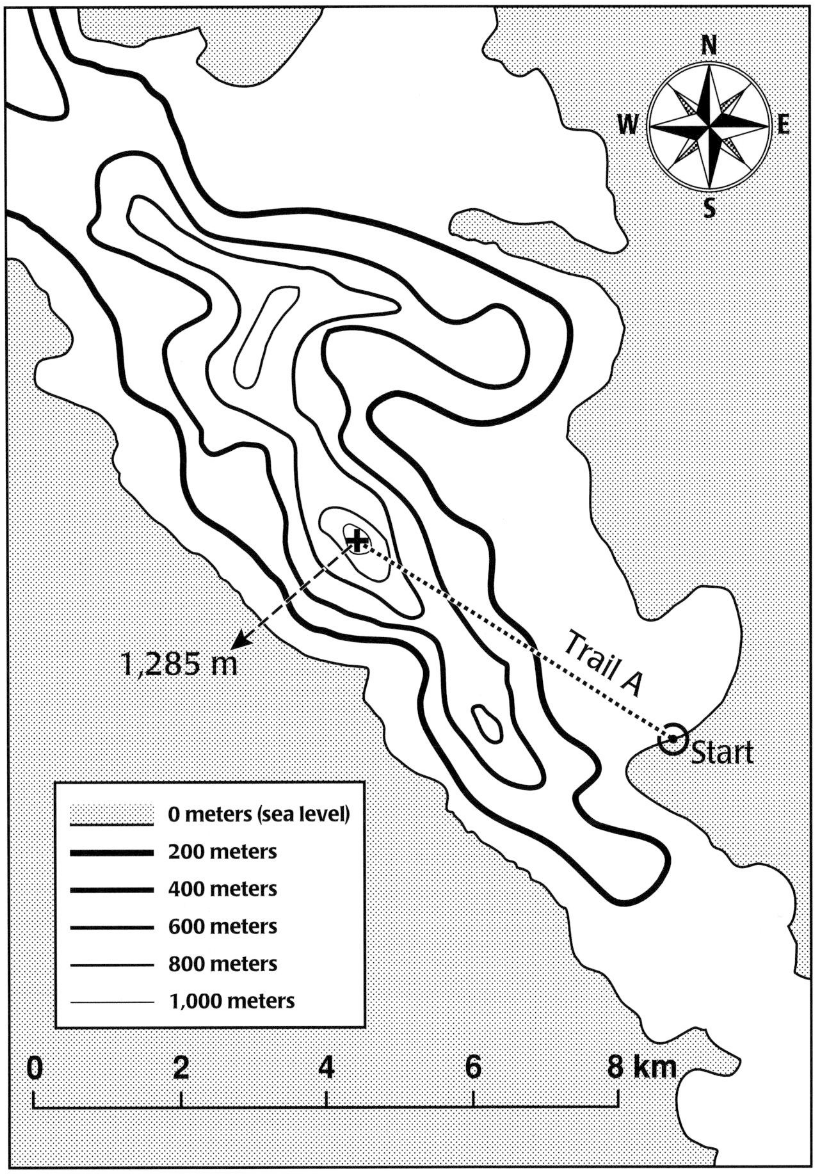

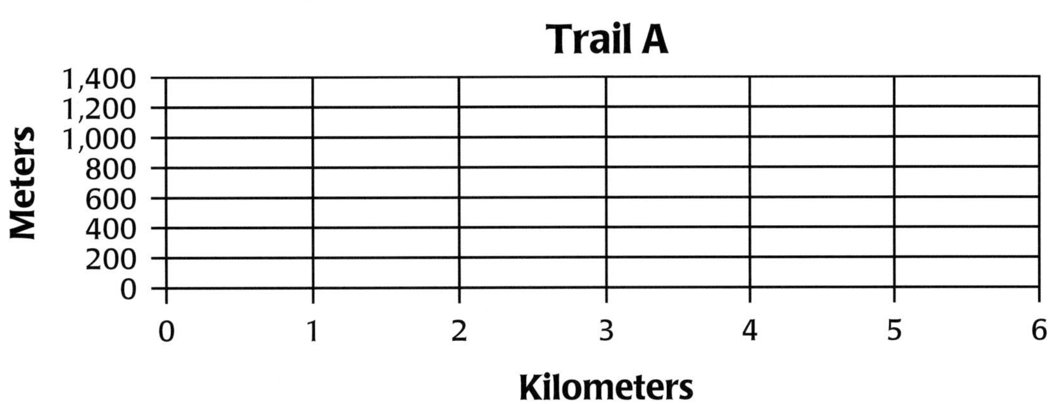

Student Activity Sheet 4

Name ______________________________

Use with *Statistics and the Environment*, pages 8 and 9.

16. **a.** Indicate the steepest part of trail B on both the profile and the contour map.

b. How can you determine the steepest part of a trail using only the contour map?

17. **a.** Indicate the part of trail B that descends on both the profile and the contour map.

b. How can you determine the part of a trail that descends using only the contour map?

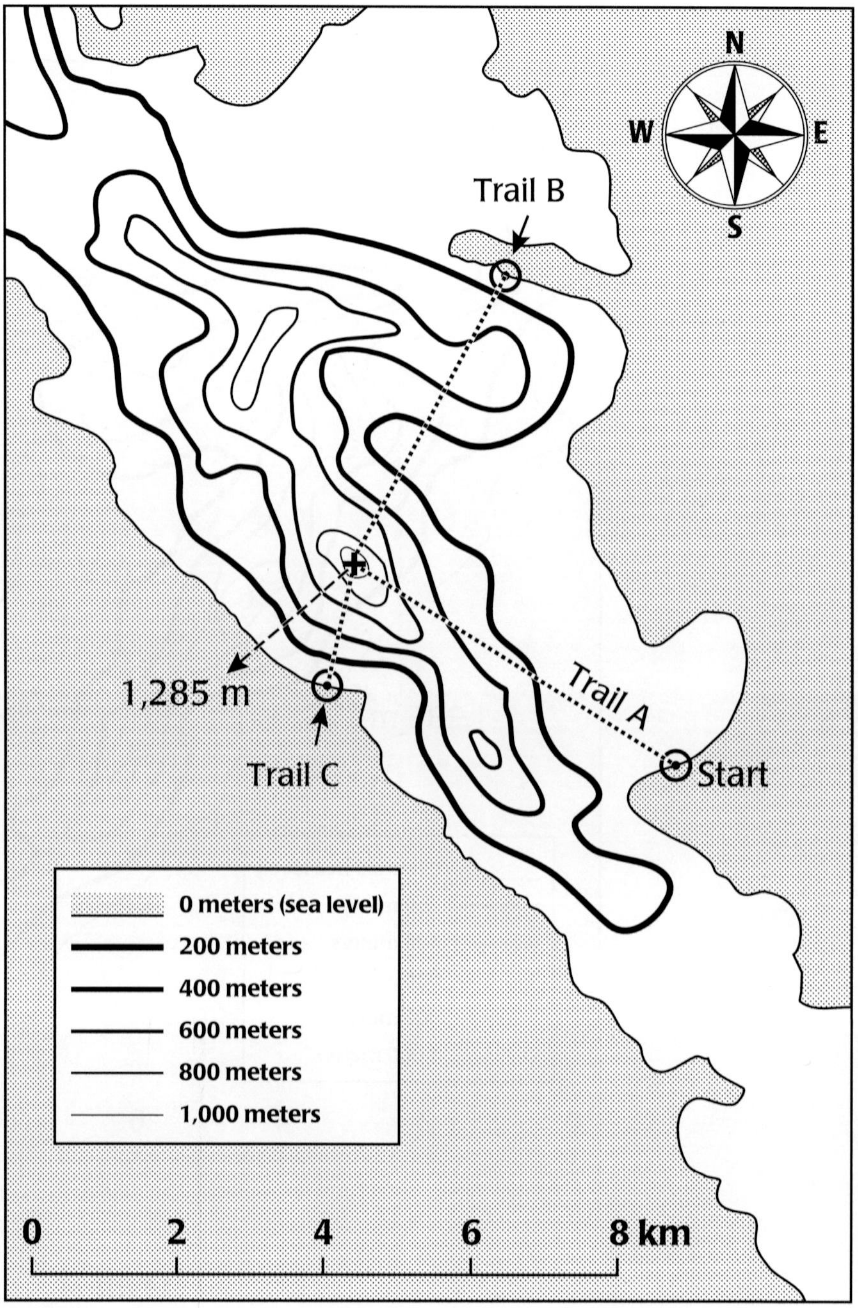

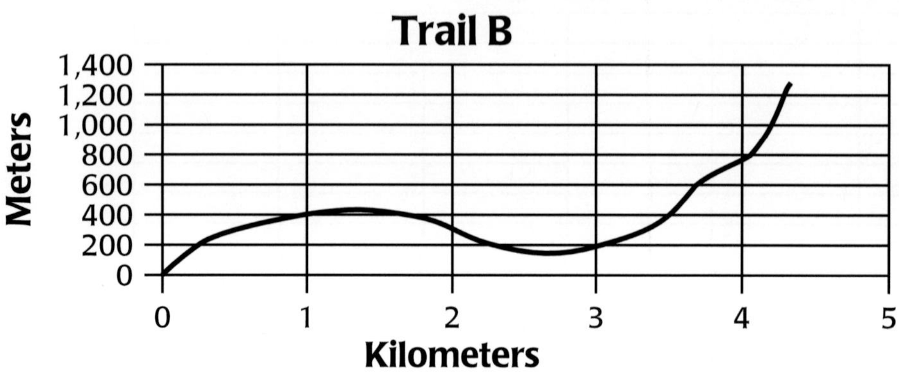

Name ______________________________

Student Activity Sheet 5

Use with *Statistics and the Environment,* page 9.

18. Draw the profile of trail C.

19. Compare the profiles of trails A, B, and C. Which of the three trails would you consider for the nature preserve on Murre Island? Explain your decision.

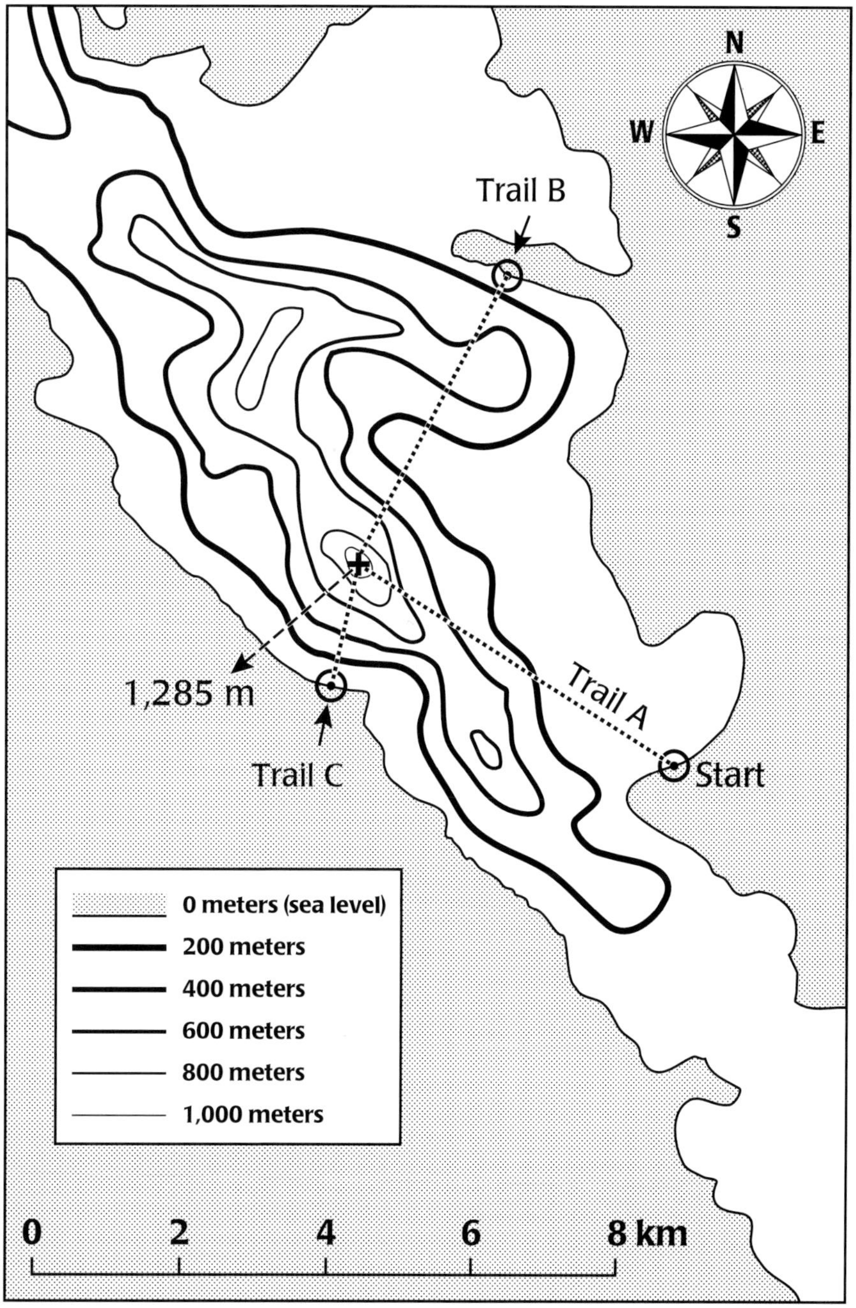

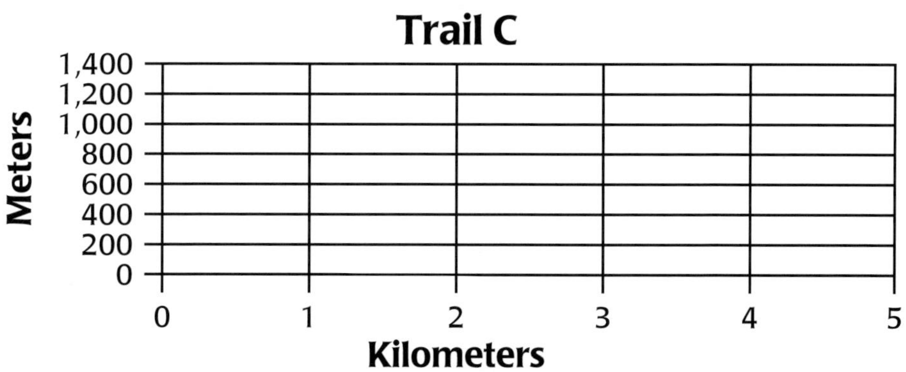

Student Activity Sheet 6

Name ______________________________

Use with *Statistics and the Environment*, page 16.

Name ______________________

Student Activity Sheet 7

Use with *Statistics and the Environment,* page 16.

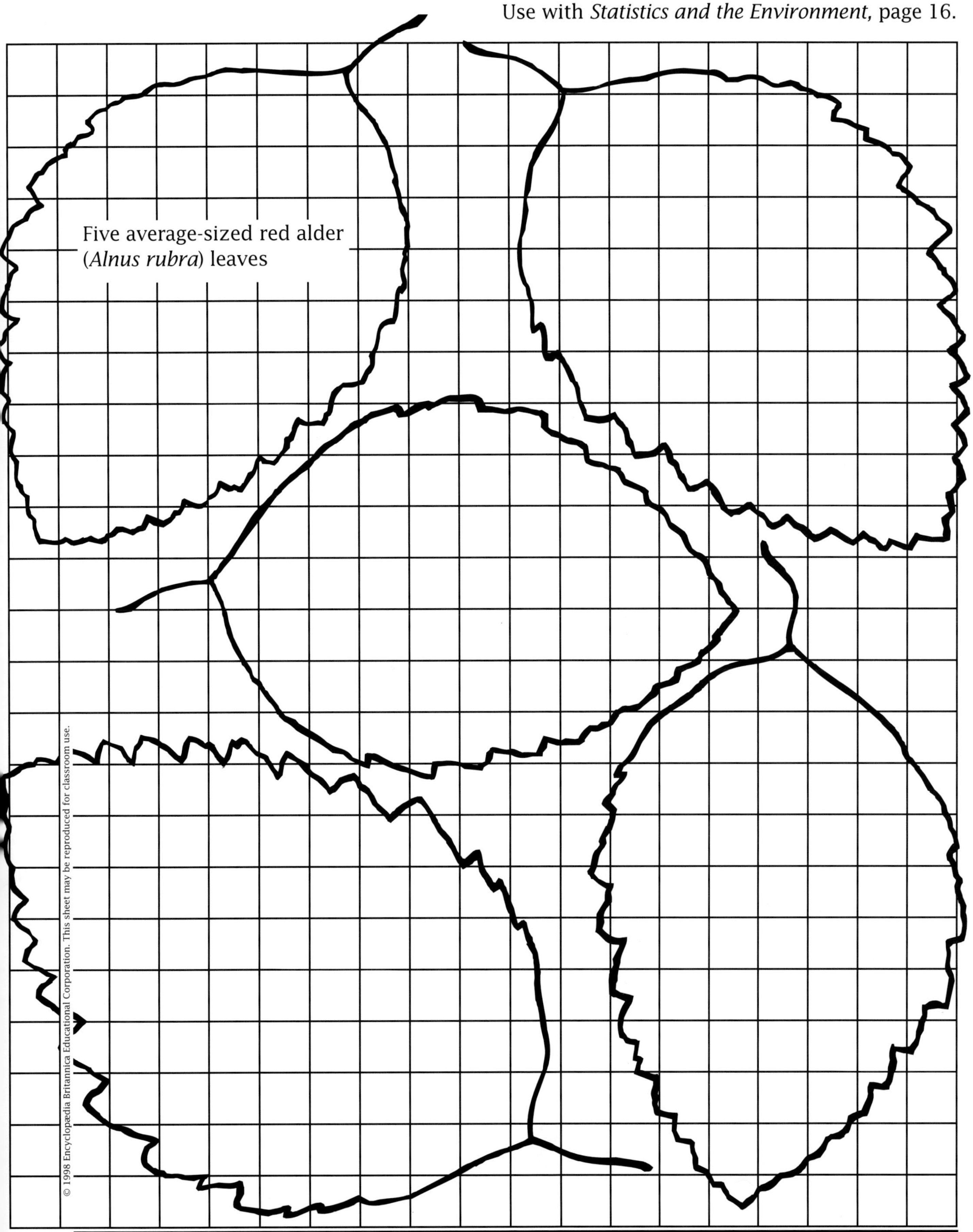

Student Activity Sheet 8

Name ______________________

Use with *Statistics and the Environment*, page 16.

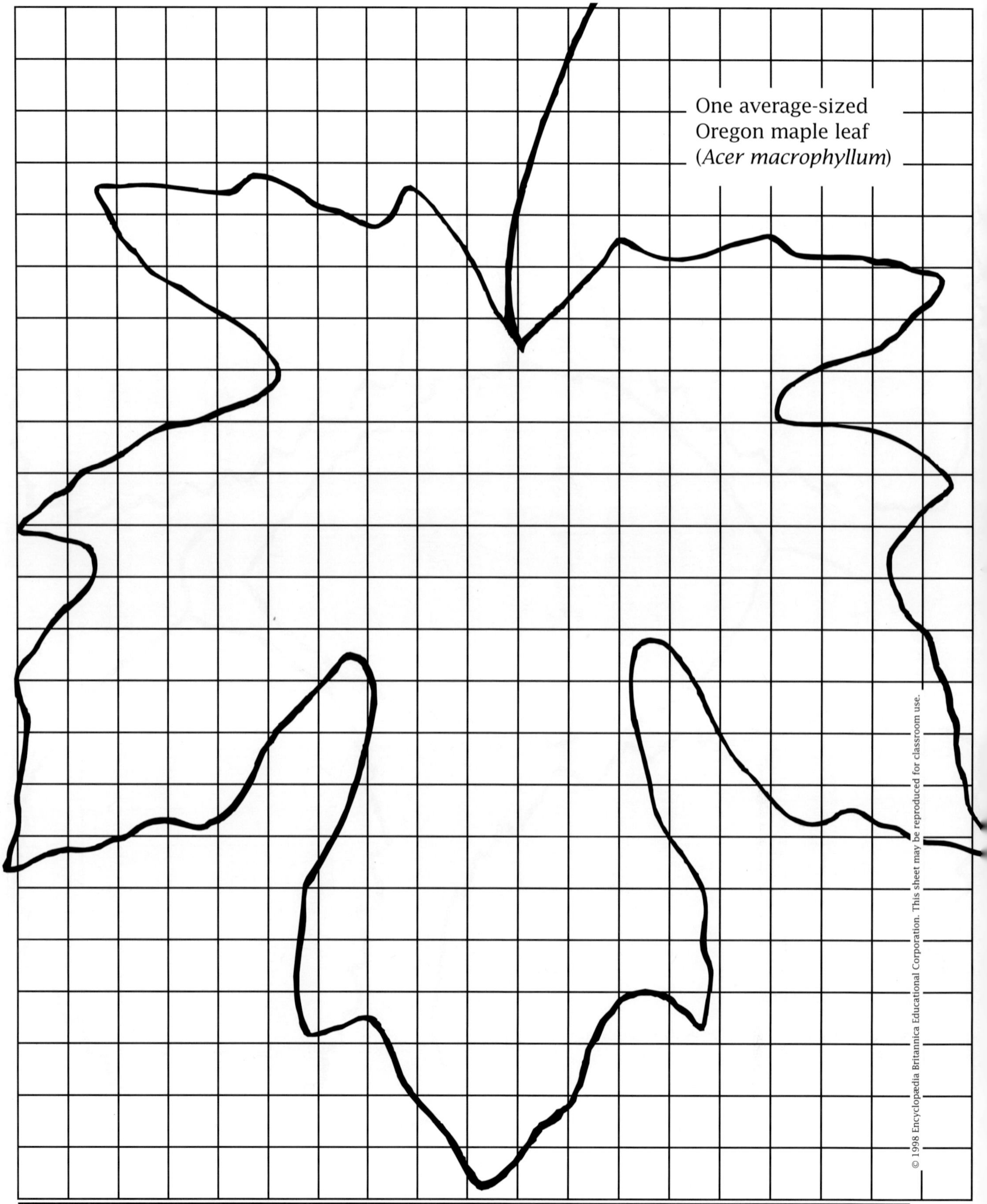

Name ______________________________

Student Activity Sheet 9

Use with *Statistics and the Environment,* page 17.

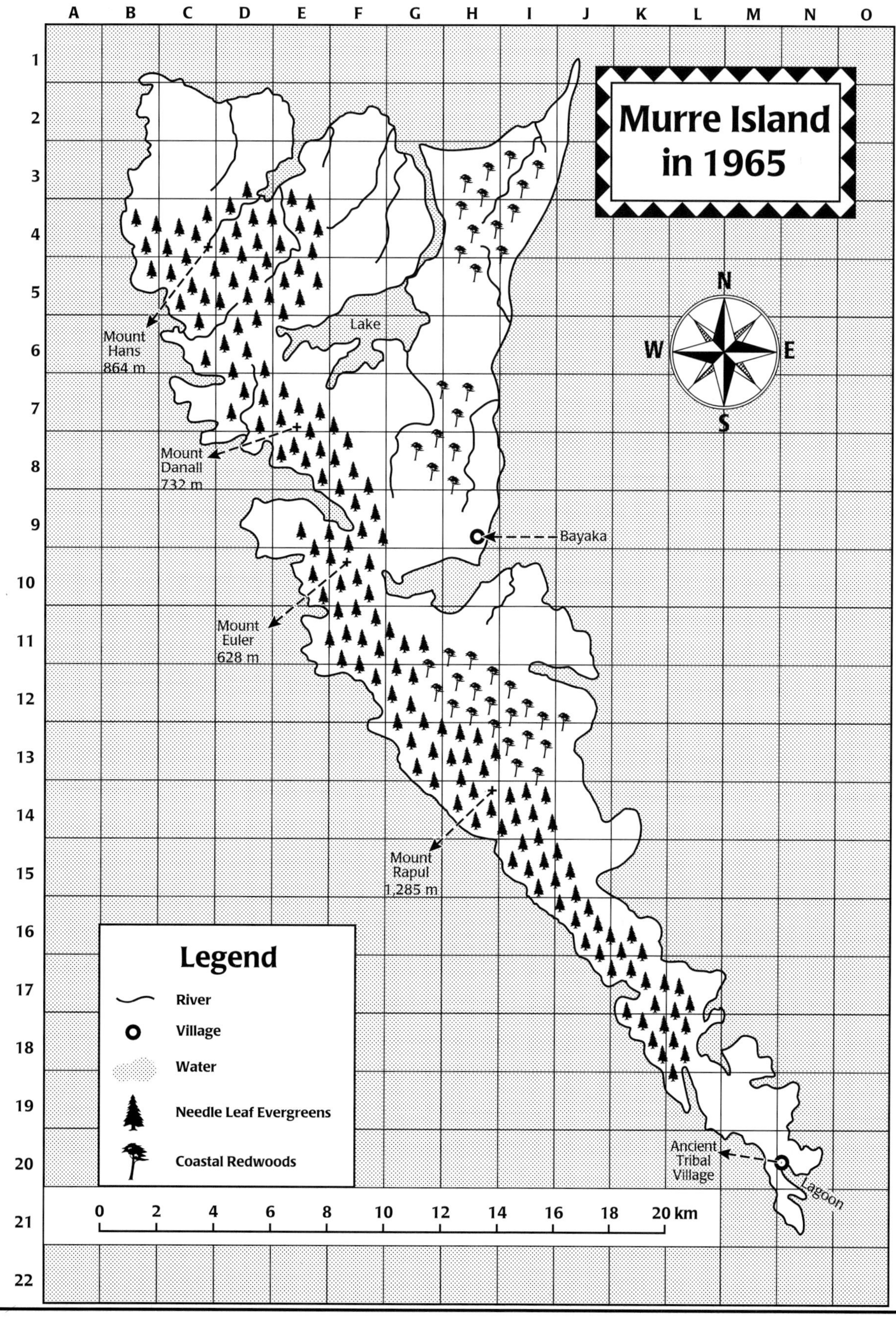

Name ______________________ Date ______________________

GARBAGE

Use additional paper as needed.

Below is a table that shows the amount of garbage per person produced in 11 different countries. Note that most countries produce less garbage than the United States.

Garbage per Person in Several Countries in 1985

Country	% Paper	% Plastics	% Glass	% Metals	% Misc.	Total (in kilograms)
Australia	26.0	6.1	15.1	7.0	45.8	680
Canada	36.5	4.7	6.6	6.6	45.7	640
France	27.5	4.5	7.5	6.5	54.0	270
Germany	17.9	5.4	9.2	3.2	64.3	320
Italy	22.3	7.2	6.2	3.1	61.6	260
Japan	38.3	7.7	1.3	1.4	51.3	340
Luxembourg	17.2	6.4	7.2	2.6	66.6	360
Portugal	19.0	3.0	3.0	3.5	71.5	220
Spain	15.0	6.0	6.0	2.5	70.5	280
The Netherlands	22.8	6.8	7.2	3.4	59.8	450
United States	34.7	6.7	9.0	8.8	40.8	740

Source: Data from The Netherlands Central Bureau of Statitistics, *Statistical Yearbook 1990 of The Netherlands* (The Netherlands: SDU Publishers, 1990), 463

1. Which country produced the most garbage per person? Which country produced the least garbage per person?

2. Use the data in the last column of the table to make a graph. You can make a pie chart, a bar graph, or a stem-and-leaf-plot. Explain your choice.

3. Which countries in the table seem to produce a lot of garbage? How is this illustrated in your graph?

Name ______________________ Date ______________________

GARBAGE

Use additional paper as needed.

Not all garbage is incinerated or dumped. Paper, plastics, glass, and metals can be recycled. Bill, a spokesman for the Environmental Society, looks at the table on the first page of this assessment and says:

> "Only 40.8% of the garbage in the United States is not recycled.
>
> The rest (paper, plastics, glass, and metal) is recycled.
>
> In all other countries, a lower percentage of garbage is recycled, so the United States produces the least amount of garbage per person!"

4. Do you agree with the above statement? Why or why not?

5. Look carefully at the data in the table, and write two statements or two problems using the data. Explain how you came up with your statements or give the answers to the problems.

Name ______________________ Date ______________________

LAKE SUPERIOR ISLAND

Use additional paper as needed.

The following journal article is about the commercial development of an island along the southern shore of Lake Superior. Carefully read the journal article.

Lake Superior Island to Be Developed

Grand Island, the largest island along the southern shore of Lake Superior, will be developed commercially as a national recreation area. The U.S. Forest Service has developed a plan for the 13,500-acre island that will include a lodge, some roads, hunting, and numerous recreational activities, but it will also leave large areas of the island undeveloped in a near-natural state. The recreational area should draw people from as far away as Madison, Wisconsin, because there are no other such recreational areas in the state of Wisconsin.

The plan is a compromise between the interests of local businesses and environmentalists. Even so, some environmentalists are still angered by the plan because they fear that the upcoming development will ruin the natural beauty of the island, which includes pristine white beaches, lush hardwood forests, and towering sandstone cliffs. They also worry about the wildlife. The 4-by-7-mile island is currently inhabited by bears, otters, coyotes, martens, foxes, eagles, hawks, grouse, songbirds, and deer.

The recreational plan includes seasonal hunting, trapping, and fishing. Campgrounds will be constructed along with hiking and biking trails. During the summer, there will be extended trails for hikers and mountain bikers. Rock climbing and jet-skiing are the only activities that will be banned at Grand Island.

1. a. List the concerns of the environmentalists who are against the development plan. Do you think these concerns are valid? Explain.

b. What other environmental issues can you think of that should be considered before the island is developed as a vacation spot?

The island developers want to know how many people to expect, and they want to know what recreational activities people are most interested in. The developers think that most people will come on a day trip only. The developers design a survey.

2. a. Explain why each of the following samples should or should not be used by the developers to complete their survey.

- a representative sample from the United States
- a representative sample from Madison, Wisconsin
- a representative sample from the states Michigan, Wisconsin, and Minnesota

b. Suggest another representative sample that the developers might use.

Name ______________________ Date ______________

HIKING ON AN ISLAND

Use additional paper as needed.

Below you see a contour map and a drawing of a small island. The numbers in the contour map indicate altitudes in meters.

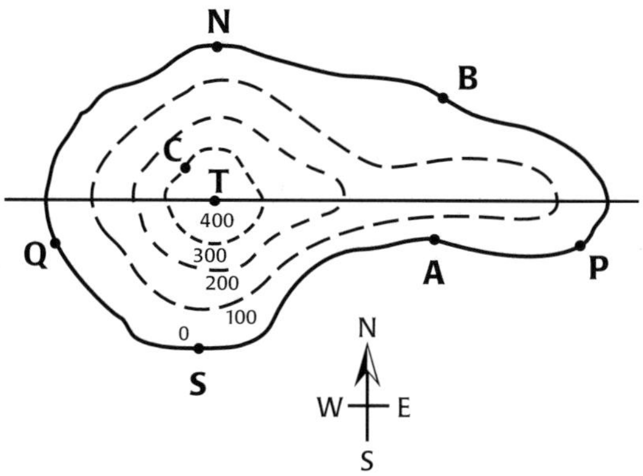

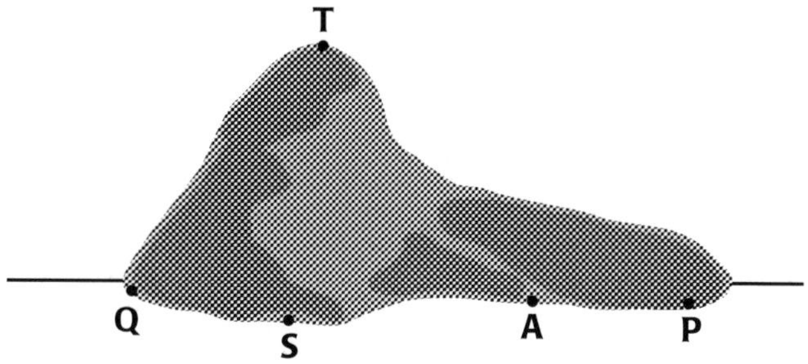

1. Draw a profile to show the varying altitudes on a hiking route that begins at the most northern point, *N,* and goes straight to the most southern point, *S.*

On the right is a profile of one hiking route on the island.

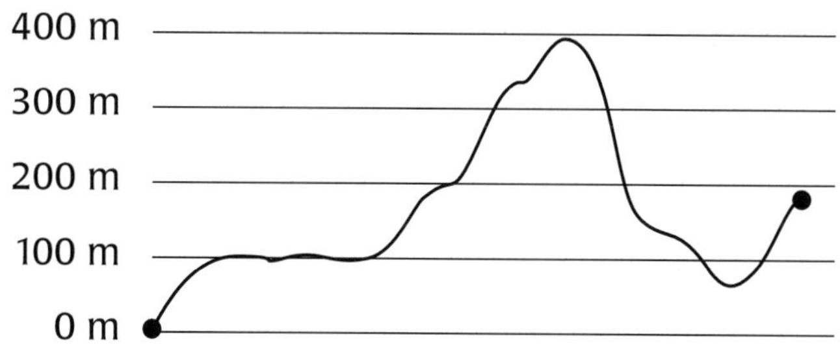

2. Describe how this route ascends and descends.

3. On the above contour map, indicate a route that matches this profile (draw in the route with your pencil).

The following are descriptions of two hiking routes from point B to point Q:

I. *From point B, cross the island, walking straight to point A.*
From point A, travel along the coastline in a westerly direction till you reach point Q.

II. *From point B, walk straight to point C, from which you can get a beautiful view of the island. From point C, walk straight to point Q.*

4. Which of the two hiking routes do you prefer. Explain.

CREDITS

Cover

Design by Ralph Paquet/Encyclopædia Britannica Educational Corporation.

Collage by Koorosh Jamalpur/KJ Graphics.

Title Page

Phil Geib/Encyclopædia Britannica Educational Corporation.

Illustrations

6, 8, 10, 12, 20, 24, 28, 29, 30, 36, 38, 40, 42, 44 Phil Geib/Encyclopædia Britannica Educational Corporation; **50, 52, 56** Paul Tucker/Encyclopædia Britannica Educational Corporation; **60** Phil Geib/ Encyclopædia Britannica Educational Corporation; **74** Paul Tucker/Encyclopædia Britannica Educational Corporation; **78, 80, 84, 88, 98** Phil Geib/Encyclopædia Britannica Educational Corporation; **100** Paul Tucker/Encyclopædia Britannica Educational Corporation; **102** Phil Geib/Encyclopædia Britannica Educational Corporation; **104, 106** Paul Tucker/Encyclopædia Britannica Educational Corporation;

Photographs

22 © Stan Osolinski/Tony Stone Images; **70** © Bruce Matheson: PHOTO/NATS; **78** © CAMERAMANN INTL.; **82** © CAMERAMANN INTL.